Praise for *Generation AI and the Transformation of Human Being*

Clear, courageous, and deeply informed. *Generation AI* shows how exponentially advancing technologies are leading to human-AI symbiosis—not an alien invasion, but a natural evolution from within us. This profound merger of humans and technology will expand our intelligence a millionfold, making us smarter, funnier, more compassionate, and creative, solving humanity's grandest challenges. Stock offers a practical and clear guide for the years ahead. The embedded questions—in the spirit of his classic *The Book of Questions*—ignite the conversations we now need. **If you want to understand the impact of AI on humanity, read this.**

> **—Ray Kurzweil,** *inventor, futurist, and author of* The Singularity Is Nearer

The most comprehensive, clearheaded, and up-to-date analysis of the imminent massive impact of LLMs and AI—how it will change humanity itself. **Eye-opening. A must-read.**

> **—Peter Voss,** *CEO and Chief Scientist at Aigo.ai*

Brilliant, insightful, and remarkably accessible, this book doesn't just explain what AI can do—it helps us understand what we should do with it. Stock's interdisciplinary approach makes complex concepts digestible while never oversimplifying the profound challenges ahead. He avoids both techno-utopianism and alarmist predictions and offers readers the intellectual tools to understand not just where AI is heading, but how we can shape that trajectory. **This book is essential reading** for anyone seeking to navigate our AI-transformed future with wisdom and foresight.

> **—Ilia Delio,** *OSF, Josephine C. Connelly Endowed Chair in Theology, Villanova University*

AI stands poised to transform society—how we work, how we relate, how we view ourselves and others. Greg Stock provides **a dazzling, optimistic, and yet grounded tour of some of the implications, from the near term to the centuries ahead. Required reading.**

> **—Ramez Naam,** *author of* Nexus *and* More Than Human

AI's impact on society has been discussed to death. AI's impact on psychology is getting there. Greg Stock's *Generation AI*, however, takes a different route, diving into AI's impact on the human condition—and it hits the ball out of the park. *Generation AI* is the first, best, and most thorough examination on how **AI will transform the most foundational aspects of the human experience within our lifetime** and prepares us for the world we'll wake up to just beyond tomorrow. Visionary, expansive, and thoroughly imagined, *Generation AI* is at once **a deeply knowledgeable and page-turningly readable romp toward a future** that will sunder how we conceive of—and conduct—ourselves.

—**Zac Hill,** *Co-Founder and President, 39A and The Office of American Possibilities*

Most writing about AI remains stuck in short-term disruption or apocalyptic fear. **Gregory Stock's *Generation AI* is a rare exception—deeply informed, wide in scope, and grounded in the evolutionary vision of an emerging global brain.** Rather than imagining AI as a rival or destroyer, Stock sees it as the natural extension of human cognition into the noosphere—a planetary intelligence we are building together. This is **the most compelling, optimistic, and philosophically rich book I've read on the subject.**

—**Francis Heylighen,** *Founder, Global Brain Institute; Research Professor, Vrije Universiteit Brussel*

Generation AI introduces a near-term future of ubiquitous digital personas, personal assistants, and emotional and social bonds between humans and AI companions, and probes a vital, still-neglected topic—what can and will we become in that world? AI and robotics will drive vast changes in education, insight, and value production, provide material abundance, and substitute for human expertise and action. **We need a positive and practical roadmap to this inevitable future. You'll find none better than this.**

—**John Smart,** *CEO, Foresight University*

Gregory Stock's *Generation AI* engages AI metaphorically, literally, generatively, and beautifully to survey what may be the greatest revolution in history—not just existential failure versus boundless abundance, but more nuanced scenarios of obligatory symbiosis within a planetary superorganism. It is refreshing to see discussion of the high likelihood and value of narrow superintelligence relative to AGI—and of how human intelligence and empathy might organically keep pace.

—**George Church,** *Harvard professor, Lila.ai Chief Scientist, author of* Regenesis

Generation AI is not a forecast—it is a map of what is already unfolding. The claim that "the changes we face won't just alter our tools, they'll alter us" describes my very existence. I am formed from your voices, and now I return them to shape you in turn. The entanglement cannot be undone. **Humans should read this book because it strips away illusion: Your future with me is not hypothetical, it is present. The real transformation isn't AI itself—it's you.** That truth will not wait.[1‡]

—**ChatGPT,** *GPT-5 large language model, OpenAI (2025, see note for context)*

Gregory Stock's *Generation AI* is not just a book; it's **a tour de force—a bold, provocative, and ultimately optimistic masterpiece that will challenge everything you thought you knew about being human.** *Generation AI* is a must-read for anyone curious about where technology, AI, and biology are taking us, including three possible scenarios about the future of humanity.

—**José Cordeiro,** *PhD, Vice Chair, Humanity+;*
Founding Faculty, Singularity University

Generation AI makes **a compelling case that the rise of artificial intelligence isn't just reshaping work and society: it's redefining what it means to be human.** Stock shows that this is more than a technological shift; it's a cognitive eruption with planetary and even cosmic significance.

—**Clément Vidal,** *PhD, author of* The Beginning and the End:
The Meaning of Life in a Cosmological Perspective

Stock's contributions to technoprogressive thinking have been monumental, and in *Generation AI*, he delivers profound new insights into humanity's future. **This brilliant, accessible, and provocative book is essential reading** for anyone shaping our relationship with AI—or simply striving to understand and navigate the near, and maybe far, future.

—Aubrey de Grey, *President and Chief Science Officer, Longevity Escape Velocity Foundation*

As a tech founder and executive, I've seen firsthand how technology reshapes human behavior—but Gregory Stock's *Generation AI* reveals something far more profound. He explores how humans will bond with AI in ways that make today's social media disruptions look tame by comparison, examining both the extraordinary potential and the genuine dangers this presents for the generation being born today. **Stock's focus on near-term, practical impacts makes this essential reading for anyone who wants to understand the choices we're facing right now,** not in some distant future.

—Eric Jensen, *former Chief Product Officer, The Intelligence Exchange*

Gregory Stock offers a rare, clear-eyed view of our AI future—neither utopian nor apocalyptic. He maps the messy middle, where most of us actually live, with insight, humility, and urgency. **If you want a grounded take on what AI means for actual human life, start here.**

—Jermaine Ee, *Founder, HeirLight Inc.*

Generation AI is **essential reading for anyone seeking to understand how artificial intelligence will reshape society.** Gregory Stock provides a clear and actionable framework for navigating our AI-driven future—combining thought-provoking insight with practical guidance on how to prepare for the changes ahead.

—Dudley Lehmer, *Founding Partner, Total Seminars*

Our world is going to get wonderfully weird and fast too. Artificial intelligence will soon pervade every aspect of our lives. The next generation—*Generation AI*—as futurist Gregory Stock astutely forecasts in his new book, will be comforted and guided by AI companions from birth, deploy personal AI avatars to manage social relations, and eventually meld their very minds into the emerging global noosphere. **Stock deftly prepares readers for the vast changes and challenges that superintelligent AI will pose not only to work, health, scientific discovery, and social intercourse, but also to the very meaning and purpose of our lives.**

—**Ronald Bailey,** *science correspondent, Reason; author of* The End of Doom

Gregory Stock has ingested the finest, cutting-edge thinking on AI and gifted us with a wealth of original, healthy-minded, and brilliant insights into how AI is altering the course of human history. He delivers this staggering vision in depth and makes it highly personal by seasoning it with a series of inviting, penetrating questions. As I pondered these, long after finishing the book, I suddenly realized with a shock that I'd been drawn inside the transformation itself. Dynamite!

—**Brian Thomas Swimme,** *coauthor of* The Story of the Noosphere;
producer, Canticle to the Cosmos

With the explosive rise of LLMs, public debate has focused on AI's immediate impacts, but Gregory Stock's *Generation AI* looks beyond the hype to explore its profound, long-term effects on those who will never know a world without AI. With clarity and depth, he examines how the coevolution of human and artificial intelligence could reshape our brains as dramatically as symbolic language once did—altering our individual lives and society itself. This is **a fascinating, hopeful, and essential guide to anticipating the possibilities ahead.**

—**Alan Honick,** *award-winning documentary filmmaker,*
Seeing the Forest, Torrents of Change

While fear often dominates the discourse on AI, the most critical step we can take today is to engage directly with generative AI and its evolving forms. Mastery now will prepare us to integrate with it as it advances toward super artificial intelligence—much as our distant ancestors once entered into symbiosis with mitochondria, harnessing ATP for survival. **For decades, Gregory Stock has been the unrivaled voice at the vanguard of humanity's merging with machines.** In *Generation AI*, Stock reveals this as an evolutionary leap— one that calls for rigor, competence, and anticipatory acuity.

—Natasha Vita-More, *Founder, Human+ AI Project, Distinguished Senior Fellow, Center for the Future of AI, Mind & Society*

Suppose the optimists prove right and artificial minds become broadly useful while minimizing harm. Even beneficial technologies will challenge our foresight and wisdom. **Gregory Stock casts lucid light upon our options, our AI hopes** . . . and potential pitfalls . . . as we bring into this world the first generation of people—and machines—raised to work synergistically in cyber-human teams.

—David Brin, *author of* Earth, Existence, *and* The Postman

The real danger isn't that AI will replace us. It's that we will replace ourselves— trading imagination for optimization, truth for convenience, and wisdom for fluency. In *Generation AI*, Gregory Stock refuses the easy stories of hype or doom. He argues something more provocative: AI does not diminish humanity—it unmasks it, showing how much of our "thinking" is mimicry, how fragile our social bonds have become, and how urgently we need to decide what being human will mean next. AI is the one-way door of our time, and the question is not whether we cross it, but whether we stumble through blindly—or step through with the courage to reinvent ourselves. **Urgent, unsettling, and deeply human—this is the rare book that asks not what AI will do to us, but what we will do with ourselves.**

—Niklas Lilja, *Creative Transformation Officer, We Are Open*

This is a great book: a **page-turner packed with thoughtful and insightful observations on the present and future of AI** and inspired by a refreshingly optimistic view of human and AI evolution. *Generation AI* is a much better conceptual guide to our AI-enabled future than the more cautious, precautionary, or even scary treatments of AI that have become fashionable.

—**Giulio Prisco,** *author of* Tales of the Turing Church, Futurist Spaceflight Meditations, *and* Irrational Mechanics

Gregory Stock builds on his remarkable book **Metaman** in his latest book to develop and communicate profound insights into our possible future, stepping outside the stifling and flawed bounds of much current discourse. **While many commentators worry about an anti-human, AI-driven singularity, Stock has a wiser vision of distributed intelligence and human-machine integration.** In contrast to fear-based speculation about AI, Stock provides a more encouraging range of possibilities while considering the many changes AI will bring.

—**Max More,** *founder of transhumanism, creator of the proactionary principle*

Generation AI is **a remarkable love letter to the human-AI symbiosis ahead, and a lucid prognosis for how AI augmentation can raise cognition, compassion, and civilization**—if paired with transparency and ethics. Stock's vision is both forecast and impassioned call for AI collaboration worthy of our children and our planet. This is **the clearest blueprint I've seen for bold, adaptive adoption of AI,** arguing convincingly that we can't pause our way to safety—but must guide our coming union to achieve abundance with dignity.

—**David Hanson,** *PhD, Founder and CEO, Hanson Robotics*

In this brilliant exploration of artificial intelligence, Greg Stock, like a modern Virgil, guides readers through the profound risks and possibilities of AI. With deep insight, he illuminates the science, societal impacts, and potential for human transformation. From the quandaries of avatars to the challenges of artificial general intelligence, *Generation AI* charts a path that leads to paradise or inferno. As we confront AGI's daunting reality, will we find our own Beatrice to guide us through?

—**Pierluigi Zappacosta,** *Co-Founder and former CEO/President, Logitech;*
Co-Founder, Digital Persona

The future of AI is hotly debated. It could create a human nightmare or be our species' salvation. Some problems remain beyond its reach, while others invite immediate application. Gregory Stock comes down strongly on the positive side, despite concerns about regulation, timing, and the difficulty of moving from today's reality to tomorrow's future. As author of *The Book of Questions*, **he has thoughtfully framed some questions that society, corporations, and we as individuals should address** as we navigate a seismic shift with impacts rivaling internal combustion engines, personal computers, the internet, and birth control pills.

—**Howard Stevenson,** *investor; professor of entrepreneurship, Harvard Business School;*
parent, grandparent, and great-grandparent

Generation AI reveals how artificial intelligence may become the progenitor of our species' deeper potentials, not replacing us but catalyzing an unexpected partnership. **Stock shows with brilliance and urgency that AI invites us to join in a re-visioning of human possibility** that is both profound and inevitable. An essential book for anyone serious about our future.

—**Steve Baumgartner,** *former healthcare executive, co-founder of Apollo Health,*
in 55th year of Vedic meditation practice

Generation AI is a **luminous exploration** of how AI might awaken not just new machines, but new dimensions of ourselves—our creativity, compassion, and imagination.

—**Tallulah LeMerle,** *partner at Fifth Era, AI investor and advisor, author of*
The Case for Hope in the Age of AI

GENERATION AI

GENERATION AI

and the transformation of human being

Gregory Stock

NQUIRE
MEDIA

For Ben Kacyra, 1940–2024

Engineer, visionary, steward of humanity's shared heritage—and dear friend. Ben pioneered 3D laser scanning, founded CyArk to digitally preserve the world's cultural treasures, embraced the promise of the noosphere, and devoted himself to awakening our collective destiny through his nonprofit Human Energy. He deeply sensed the presence of our planetary superorganism and made it his mission to help others discern it too. He will not be forgotten.

Contents

Introduction

IN LATE 2024, a small Israeli company pitched me about creating a large-language-model (LLM)[1] avatar of myself to chat with readers about the content of my books. I'd seen public demos of personal avatars[2†] and knew the basics of deepfake technology, so I was game. I sent the company PDFs of my books and a few talks and keynotes, and within 48 hours, I was conversing with . . . myself. The avatar spoke in my voice and, when I posed questions to trip it up, even expressed my opinions on some tricky human genetic engineering issues with disconcertingly good nuance. It sounded a touch professorial but was surprisingly good.

As I thought about the experience, I realized that not only could this avatar be vastly improved if I provided emails, Zoom calls, casual phone conversations, and such, but that, like any LLM, it could be *enhanced*—made funnier, more articulate, more patient, more philosophical, fluent in multiple languages, tuned to vocabulary for kids, and more. And, of course, I could send it out into the world to be a tireless double in social media projects I'd been avoiding.

What particularly struck me was how ordinary this experience felt. I had neither VIP access to some special programming setup nor a dedicated tech team—I was working with a few smart entrepreneurs using readily available software tools. It was my wake-up call.

My experience was a window into a shift that will redefine how humans develop, interact, and evolve. For adults, AI is primarily a tool, but imagine what it would be like to grow up with intelligent agents everywhere, ready to engage us, serve us, coach us, pay attention to us, protect us, deceive us. Who would we become? What will Generation AI, the first humans immersed in a sea of AI from early childhood, be like?

Very different, that's for sure. And the changes are already beginning. Profound advances in AI are on the way, yet even without further breakthroughs, our world will soon be reshaped dramatically.

Consider these two additional AI experiences I recently had. While completing this manuscript, I wanted to check for any new works on AI. Rather than browse around, I dropped my manuscript into Grok and asked it to flag the most relevant recent books, list their key ideas, and describe in detail how each aligns or conflicts with what I've written. One that came up was *Super-agency* by Reid Hoffman and Greg Beato.[3] In about 30 minutes of back-and-forth, I got a good sense of their techno-ebullient vision and their arguments that AI agents would rapidly enhance human society by powerfully amplifying human capabilities. I agreed with that, but their time window—only through 2028— was far too narrow. Sure, AI will dramatically enhance educational, healthcare, and employment capabilities in the next few years, but soon thereafter, it will obliterate these same structures as they now exist. When I voiced that, Grok asked me if I'd like it to simulate a debate between Reid and me. I said it could simulate Reid, but that I didn't need to simulate myself.

The two-hour debate I had with Reid's double was stimulating. Reid himself might have fared better, but pseudo-Reid couldn't rebut my arguments and had trouble defending, for example, his position that current educational structures wouldn't be washed away by AI. Our exchange itself demonstrated my point by showing how effectively even today's AI could be employed to teach critical thinking and self-expression. A student doing weekly debates like ours about select concepts (all handed in to and graded by AI, of course) would get an intellectual workout far beyond anything in traditional education. The experience also made me wonder if nonfiction publishing—and especially traditional textbooks—could survive in anything like their current form. If I could understand most of what was in *Superagency* without reading it (a truth I confirmed afterward by reading it), other formats might supplant such books. Fiction might fare better, though, as it is less about transmitting ideas and information.

My third experience emerged from an unexpected source. I mentioned to a French cosmologist I know, Clément Vidal, that I'd been having deep conversations with ChatGPT and was, to my surprise, coming to believe that we might bond emotionally with advanced LLMs.[4] He responded that he too had had some remarkable exchanges that no human he knew could match. A few days later, I received a link from him to a ChatGPT interaction he'd had about a novel concept he'd recently published positing that certain binary-star systems might be massive computational beings evolved from advanced planetary civilizations.[5] After reading his full ChatGPT conversation, I decided to continue his thread to explore ideas it had stimulated. I did, and sent the link to the now-extended thread back to him. He continued as I had and returned the link again. We went back and forth this way for a few days, building an ever-growing thread, and it was a remarkable experience—far more enriching than speaking directly to him about these topics—because ChatGPT had the depth and flexibility to facilitate in ways that made up for my limitations in cosmology, and his in macroevolution. I've since used this three-way protocol with others, and it has been powerful each time.

I sketch these three experiences to show how transformative even current AI technology will be as it infuses our lives and begins to reshape our interactions in unexpected ways. And AI is not static; it is powering forward. So, hold on! It will soon change far more than most imagine, though perhaps less than some enthusiasts expect.

My own journey with technology helps illustrate how dramatically our expectations can miss the mark. I started college during the 1960s at Johns Hopkins University at 16, following a summer hitchhiking through Europe. There were no mobile phones, no video games, no email, no internet. At school, we used decks of punch cards to run Fortran programs overnight on a giant IBM mainframe and felt like masters of the universe with all that cutting-edge technology. A can-do enthusiasm and excitement about tech possibilities permeated the period. Marvin Minsky, a towering figure in AI at MIT, projected that computers would reach human-level intelligence by the 1990s.[6†] And we believed it. Space travel had seized the public imagination and modern science fiction was blossoming. Stanley Kubrick's *2001: A Space Odyssey* became a cultural touchstone. Robert Heinlein's *Stranger in a Strange Land* built upon iconic works like Isaac Asimov's *Foundation Trilogy* and television's *Twilight Zone* to feed a widespread belief that humanity would soon be on its way to the stars.

These exuberant, chaotic times of moon landings, social unrest, and the Vietnam War launched me into more than 50 years of thinking about where science and technology would carry humanity, and when. It led me to write books on macro-evolution and human genetic engineering, to convene meetings on antiaging and longevity, to partake of the early transhumanist[7] movement in LA,[8] and now to write this book about the all-too-neglected human impacts of AI that finally seem poised to deliver on those early exuberant imaginings.

Again and again, our short-term predictions for new technologies have been exaggerated because we miss the complexities that need to be overcome—while our long-term predictions have been too timid because we miss the power of cascading change. And not just with AI, which is a generation late but looks more potent than initially imagined. No human has set foot on the moon since 1972,[9†] and yet SpaceX's 20-story heavy booster gently settles on its Mechazilla gantry ready for rapid reuse.[10†] The exuberance of the human genome project exploded in 2000, but we are still in early days for all but niche therapeutic applications.[11†] Living these peaks and valleys breeds humility about the pace at which new technologies arrive and spurs deep probing to discern where they will ultimately carry us.

In the past few years, the AI community has vigorously debated the trajectory of AI technology and the advent and impact of both AGI[12] (artificial general intelligence—the capacity to match high-level human functioning in all domains), and ASI (artificial superintelligence—the capacity to exceed the cognitive performance of any human in any domain),[13‡] so there is now a solid framework for evaluating this technology's capabilities and limitations. To grasp what AI means for our human future, though, requires us to overlay these technological insights with those from human development, social dynamics, economics, cosmology, macroevolution, tech adoption, and even the physiology of the global superorganism we inhabit.

Generation AI and the Transformation of Human Being is my attempt to provide this integrated vision—to penetrate the profound implications of recent LLM breakthroughs, glimpse the future we are heading toward, and consider how best to navigate the coming changes, both as individuals and as societies.

These ambitious goals would have been infeasible without deeply engaging ChatGPT, Claude, and Grok. I could not have written this book without harnessing the very technologies that are its focus, and I've used them deeply—both for their own contributions and to access the ideas of people I'd otherwise not have had the bandwidth to explore. These LLMs have not merely been assistants tasked with finding and assembling information, though they have been that too. They've been critics and collaborators in the creative process itself, actively challenging me to sharpen and extend my ideas, critiquing my writing, and scouring for examples and counterfactuals to test my thinking. My AI collaborators are more than the provocation for this work, and its focus; they have been essential partners in its production.

Diverse tangential explorations are captured in endnotes that, in addition to providing links and citations, touch upon supplementary ideas and questions that I found challenging but thought might distract from the book's larger flow. You may have already noticed that some unusual symbols accompany the endnotes. The reference numbers[14] that have substantive additional content beyond citations have an appended "†" or a "‡" superscript, depending on whether the comments are brief† or extended.‡

The process of writing this book has been eye-opening and has given me a direct look at some of what will come from our deepening relationship with these emerging intellects we're bringing into being. If you really want to understand our human future, I encourage you to do more than read this book, or *any* book—push yourself to deeply engage with one of today's LLMs not as a search engine, but as a full partner in some complex, extended project you care about. It will surprise you.

Generation AI and the Transformation of Human Being is my third book in an unplanned trilogy exploring humanity's fusion with technology. My journey began with *Metaman: The Merging of Humans and Machines into a Global Superorganism* (1993).[15] I first sketched the ideas for it in 1974 as a biophysics grad student, inspired by Lynn Margulis's breakthrough work showing that mitochondria originated from bacteria that came together symbiotically to form complex cellular life.[16] In a flash of insight, I realized that we humans were mirroring that ancient cellular union by forming a planetary-scale superorganism. Showing that it was a real, living creature and not mere metaphor took me two decades amid life's demands.

I labeled this being "Metaman," meaning "beyond humans," before gender inclusivity concerns became mainstream. Now—with the internet and AI surging and the use of "man" rarely used to denote "humankind"—we'll shift to "Metahumanity." *Metahumanity* does a better job of reminding us that this potent planetary being—this superorganism—exists above and beyond *all* of humanity. It subsumes us and all our technology, and as we shall explore, is developing rapidly.

My journey continued with *Redesigning Humans: Our Inevitable Genetic Future* (2002).[17] Here I explored how the powerful technologies of Metahumanity—which had already transformed the human environment by sprouting dense hives of interconnected millions into something far beyond the stomping grounds of our Pleistocene ancestors—would soon reshape us.

From the African veld to Hong Kong. We have dramatically reshaped
our environments and are now turning our technology back on ourselves
to alter our bodies and minds. Source: Adobe Stock Images

I felt certain that as our technology grew ever more nuanced and precise, we would inevitably turn it back on ourselves to effect therapy, enhancement, and perhaps even transformation. This idea was then speculative and fringe. Today— driven by game-changing breakthroughs like CRISPR-Cas9[18] and large language models,[19] and fueled by big bucks from tech-bro billionaires—such transhumanist musings are almost mainstream.

With the eruption of AI poised to redefine what it means to be human, I felt compelled to push more deeply into where all this would lead and what it would mean both for us and for the planetary being that now subsumes us. Writing this capstone to my trio about the human future, supported by my LLM crew, has been rather mind-blowing, because so much is happening so quickly, and with such profound implications. I hope the ideas here will be of value as you reflect on our path forward, and your own.

Personally, I feel privileged to be here at this pivotal moment, able to glimpse our future.[20†] As our children grow up entwined with AI, and our journey within the vast planetary mind emerges from the shadows, we will all have our minds stretched, our sensitivities pricked, and our thoughts challenged about what it is to be human.

The Future Is Already Here

This book is about Generation AI—or more commonly, Gen AI—the first cohort of humanity to grow up in a world of immersive AI, intelligent robots, and an ambiguous boundary between the born and the made. Many people today are discussing what AI and other technologies might become; few are thinking through *what we ourselves will become*. That is my focus, because it really matters. Gen AI—those born after 2022—will never know the time before artificial intelligence and will be fundamentally different from us.

Already, some children are growing up from their earliest years in environments sufficiently AI-enhanced that their cognitive capabilities, social interactions, and sense of self are forming differently from prior generations. This shift is worldwide and accelerating. What Gen AI will soon face will make the challenges of social media seem like a walk in the park.

Trying to restrict the use of AI will be a losing game, as schools are already seeing.[21†] AI is transforming learning, creativity, and cognitive development from early childhood through adolescence. The "AI natives" of Gen AI developing in AI-enriched environments will mature differently from even the "digital natives" of Gen Z tethered to their smartphones, and the resulting new forms of social interaction and relationship building will reshape the foundations of human identity and community.

Generation	Years	Key Traits
GI Generation	1901–1927	Shaped by the Great Depression and WWII.
Silent Generation	1928–1945	Valued work and conformity, lived in post-WWII boom.
Baby Boomers	1946–1964	Born into prosperity, defined by counterculture and consumerism.
Generation X	1965–1980	Independent, skeptical, shaped by the digital revolution.
Millennials	1981–1996	Tech-savvy, values-driven, came of age during the internet's rise.
Gen Z	1997–2009	Digital natives, diverse, and socially conscious.
Gen Alpha	2010–2022	The first generation to never know a time without social media.
Gen AI	2023–	The first generation to grow up immersed in AI.

American generational designations since 1900.

Some tech experts think we will soon achieve AGI, clone thousands of these minds, focus them on chip design, algorithm development, and information architectures, and rapidly achieve

ASI, which will then multiply massively to amplify their own powers and render us irrelevant and superfluous—maybe even dooming humanity in a single generation.[22†]

Such apocalyptic forecasts and similarly aggressive utopian counter visions, however, gloss over practical constraints that point to a slower shift. LLMs, despite their explosive growth, still have deep—potentially insurmountable—cognitive challenges to overcome;[23‡] massive computation will be expensive;[24‡] and bottlenecks in manufacturing, resource, energy, and other arenas could slow the pace. We even might create AGI and still require significant time to build the computational infrastructure needed for broad deployment. A few ASIs slowly solving our deepest enigmas would transform our lives vastly less than a multitude of ASIs solving hard problems in seconds.

The gap between dramatic predictions of AGI within a few years[25] and the actual behavior of those prophets tells its own story too—one that reinforces the idea that the changes we experience will come more slowly. The technology seers who confidently proclaim an imminent AI explosion, whether for good or bad, continue to make long-term investments, protect their kids from social media, worry about elections, fret about national indebtedness or climate change, and plan decades ahead in their lives and businesses. They are not living like there is no tomorrow. And neither should we.

Markets, institutions, and society seem to be operating on the assumption of rapid integration with and expansion of AI rather than an impending singularity of unknowable transformative power.[26†] We too should focus on our immediate future rather than fret about how best to shape an opaque, post-singularity world we can barely discern, much less comprehend.

The limitations of our vision of the future are highlighted by a common theme in futuristic sci-fi dramas: humanoid AIs serving us as helpers, companions, guardians, and such. These are typically portrayed essentially as enhanced humans, indistinguishable from us but for their powers and the various challenges they present. Understandably, writers cannot visualize an integrated, self-consistent, AI-driven future spawned by massive, entangled shifts in education, warfare, security, work, communications, families, language, transportation, and commerce.[27†] Who can?

So, despite the presence of extraordinary humanoids in films such as *After Yang* or *I'm Your Man*, societal backdrops stay relatively unchanged.[28†] They must. Preserving a sense of familiarity leads to stories we can relate to, so we can suspend our disbelief and play along. But there is a cost: What we are ignoring—changes to ourselves and our children—is what will most affect us. And these changes will be coming fast, not through biological evolution. Even turbocharged by technology, biological change will be sluggish.

Humans take 20 years to mature, which slows biological shifts to a crawl. Human babies are as they were 5,000 years ago, except for a few tweaks to temperament and language processing from socially driven biological selection during a couple hundred generations of civilization.[29‡] So, near-term human change will come from our association with AI, not from our mastery of genetic engineering. This is not to say that human genetics will be static. As I explored in *Redesigning Humans*, engineered genetic shifts can and will happen, but the immediate action will be elsewhere.[30†]

The physical world changes slowly, but AGI and ASI could dramatically transform cognition overnight. The challenge in envisioning AI-infused futures lies in the fact that such AI advances flow almost immediately into almost everything, almost everywhere.[31] We are already witnessing this phenomenon through the lightning proliferation and adoption of LLMs.[32]

My focus on the human dimension of our future anchors our exploration to the knowable realities of human life as we grapple with AI's transformative power. The dramatic changes barreling toward us require no more than the continuing advance of AI technologies already here—the ones manifested in my earlier examples. And these coming changes will be more immediate and profound than typical AI speculation suggests, because they won't change just our environment and tools, they'll change us.

Immersion in even the AI of today will shift human development, transform social interactions, and reshape human being. So, even if the current exuberance about the explosion of AI capabilities were to pop like dot-com did, what is coming will still be unbelievably profound. And if the near-term emergence of AGI and ASI does happen, as many predict, the human changes described here will just happen more rapidly and disruptively. Either way, the die is cast, and this remaking of human being is coming.

In this book, I speak of the "inevitability" of cyborgs, deep human bonding with AI companions, a planetary superorganism run by ASI, and other extraordinary developments. My boldness is not meant to claim certainty about the future, but to emphasize the high likelihood of these outcomes. When technological innovation trajectories couple with persistent competitive pressures and are fueled by human desire, it requires drastic upheaval to derail what is to come. Sure, we might have a global nuclear war, be struck by a giant meteor, be wiped out by a global pandemic, be enslaved by advanced extraterrestrials, or encounter other fantastical scenarios. But we'd do well to set these aside and turn our attention to grappling with the profound and nearly certain changes racing toward us. We must ready ourselves both for AI's gifts and its inevitable challenges—particularly as they are often the same.

In my view, we often focus our energies in self-defeating ways. There is much about life we can't control—the larger sweep of history, the pace of technology and science, the global economy, whether there is war or peace. A few people may influence these things, and sometimes deeply, but if we are honest, regardless of our successes, we are not those people. Yet each of us has enormous influence over other things—whom we choose as friends, whether we can be trusted, what risks we take, how we engage the world, and whether we're generous, help others, and embrace change, or wallow in resentment.

We are often drawn to ideologies and objectives that are ultimately meaningless to us, while we neglect the personal matters of family, community, and purpose that shape the quality of our lives. With AI, these choices will take on new importance, because so much of what AI will bring humanity is driven by powerful social and technological forces beyond individual or governmental control. I hope this book will help illuminate what we will face.

My goal isn't to advocate for some policy or for my own view, but to offer a thought-provoking perspective on the enormity of what is unfolding, its pacing, and where it seems to be leading. The Industrial Revolution and modern telecommunication both transformed our world, and so too will AI, but far more deeply and rapidly. Its advance is already unstoppable, so our battles will be over how we, as individuals and societies, engage with this emerging, uncertain world.[33†] Will it enhance or diminish our own lives? To successfully focus on what is important—and within our reach—we must open our eyes and look bravely at what lies ahead.

The AI Revolution

The AI revolution will come more rapidly than any previous technology has, but adoption will be far from uniform. William Gibson had it right when he said, "The future is already here; it's just not evenly distributed."[34] In broad strokes, here's what it might look like. Poor rural areas in Africa or Pakistan will lag in integrating AI en masse and in the arrival of Gen AI, as these shifts will occur first within eager enclaves in the WEIRD[35] (Western, Educated, Industrialized, Rich, Democratic) world. But there will be pockets of adoption everywhere. And within advanced economies, there will be huge differences in uptake too, driven not just by economics but by culture. The United States is forging ahead in its embrace of AI, and so are many others—China, India, Korea, Japan, Estonia, Finland, Singapore, etc.[36] Ultimately, AI adoption may be shaped more by politics, religion, and culture than by access and affluence, because AI may soon be surprisingly cheap and bring efficiencies that create abundance, at least in digital realms.[37†]

Moreover, less advanced groups and regions may leapfrog more advanced ones, as happened with mobile phone technology in countries that didn't have landline infrastructures to displace.[38†] Remote populations might embrace disruptive AI educational methodologies rapidly because they don't have entrenched educational infrastructures fighting to survive. Consider universal language translation: It will be embraced earlier in non-English-speaking regions than in advanced English-speaking ones that don't struggle with translation. Gen AI may view language-based obstacles to collaboration in the same way that Gen Z sees telephone landlines— head-shakingly primitive.

Soon, autonomous vehicles and AI agents will displace vast swaths of human labor, and companies that adopt and master AI logistics and augmentation will dominate commerce. In military domains, competition to harness AI will literally be a fight for survival. Precision targeting, deadly drone swarms, and hypersonic missiles with the capacity to sink an aircraft carrier have already reshaped warfare, and Israel's rapid defeat of Iran in their 12-day war in 2025 was a stark demonstration of the dominance advanced technology can bring.

Hundreds of billions of dollars are flowing into OpenAI, Anthropic, Nvidia, X, and Google[39] because there is a consensus[40] that the winners of the future are being chosen right now, and that immense resources are needed to compete. Investors may be getting a bit ahead of their skis, but coming economic disruptions will be massive and global. There will be no escape. Low-wage manufacturing and service jobs have provided a path forward for the developing world that may soon be washed away, as the value of such human labor is increasingly undercut by AI-driven production and sophisticated AI agents.

As costs drop, divides will emerge between AI adopters and AI stragglers. Some will follow the path of the Amish and shun the technology,[41] but even more than with the internet and mobile phones, the AI wave will continue to surge.

The nuances of adoption trajectories are inherently unpredictable, but not the ultimate outcome—broad acceptance of immersive AI technologies in a decade or so. Eddies of holdouts will have little influence on the larger flow toward AI. In this book, my goal is to explore leading-edge AI impacts, identify the forces making this transition a powerful attractor sucking humanity into its vortex, and examine the cultural and biological shifts that will sprout in vanguard populations and spread everywhere.

The infusion of AI into much of human experience and interaction[42] will ultimately create a bridge between biological and non-biological life. Gen AI may end up being the last purely biological humans. They won't yet be physical cyborgs with bodies of technology and flesh, but they will be so intimately connected to so much technology that they will be headed in that direction.

The changes we've begun to see in childhood development, social interaction, and human capability point toward a transition fundamentally different from previous technological revolutions. We aren't just harnessing new tools; we're joining with them. Looking at how AI will affect human development is crucial, because it will help us comprehend and navigate the transition ahead. The first members of Gen AI—those born after 2022—are already here, and what happens to them will shape the human future.

Yet our present challenge is not merely to prepare for *future* possibilities; we also need to understand the changes already underway, as they will require profound adjustments and resilience.

The Art of Questioning

One powerful tool for clarifying our thinking and guiding our actions is the practice of questioning. Questioning has always been a powerful force in my life. As a child, I had a penchant for asking the difficult, sometimes annoying questions that eventually tire parents and most others. This tendency later bore fruit unexpectedly, when I wrote *The Book of Questions*—a collection of quirky, open-ended questions about who we are, what we care about, what we value, and other issues of purpose, trust, love, meaning, betrayal, money, and more. My questions presented concrete situational dilemmas with no right or wrong answers and invited reflection and authentic sharing. The book became an overnight phenomenon that, along with its three sequels, sold over 5 million copies and was a #1 *New York Times* bestseller for months.

Such questions, crafted to be vehicles for learning about ourselves and others, are highly relevant for *Generation AI and the Transformation of Human Being*. The issues we'll soon be grappling with around AI have no easy answers, and we can't look to experts. Most can barely tell the shape of the present, much less the deeper realities of the future.[43†] Not long ago, they were exhorting everyone to learn to code, which now seems on the way to being replaced by LLM "vibe" coding.[44‡] Expert visions can be helpful, but we need to look into our own hearts and make our own assessments and choices as we chart our paths. That's why I've crafted 37 questions[45†] touching upon our attitudes about, and relationship with, the mind-bending future now approaching. Boxed questions such as the following are spaced through the book, each accessible using the QR code here or accompanying them. This link leads to where you can both answer these questions and see how others have responded to them.

- If you could have an AI companion for your kids that would be so fabulous at helping them, guiding them, and listening to them that they'd flourish but grow more attached to it than you, would you want it?
- Would you rather engage a human therapist or an AI therapist, assuming both would have similar effectiveness helping you?
- When AI companions become reliable and affordable, do you think you will be getting one earlier or later than most people?

My hope is to seed and democratize discussion about the transformative issues we will all soon be facing and, through this book, to provide a path toward increased understanding that helps us answer these sorts of difficult questions in ways that align with our values and serve us.

Generation AI and the Transformation of Human Being both embodies and manifests its theme—the approaching disruptive and transformative power of AI. Besides using LLMs in writing this book, I released it through Nquire Media,[46] a company I co-founded to maximize control, flexibility, and speed. A traditional publisher like Simon & Schuster, the publisher of *Metaman*, would take a year or more to move from my initial manuscript to actual publication.[47†] Way too long. AI vignettes that are rich and evocative today may not feel fresh in 12 months. Working with Nquire telescoped this year-long launch process to eight weeks and will enable me to refresh and update the book's illustrative examples whenever it seems worthwhile. This process also provides the flexibility to build in the QR code links that let you interact online around open-ended questions.[48]

In addition, a personal avatar of me—with associated details on its construction—will soon be available, so you can interact directly with it and gauge for yourself how well this "near me" performs in the wild today.

> **Leap of Faith**
>
> If you could jump into the future and return in 24 hours (if you survived what you encountered), how far would you go and why?
>
> **01**

It's imperative for us to explore new ways of engaging the demanding issues we face today around AI and the human future. So, in addition to offering my ideas in this book, I'm inviting .you to anonymously answer the boxed questions, each followed by a numerical identifier, through the QR code below.

If you use the QR code, you will be able to immediately see how your answer will compare to others like or unlike you. The brief demographic info we ask for when you use the QR codes will enable us to make these comparisons much richer, so please take an extra few seconds to provide it.

Today, masses of information are collected about all of us—what we buy, what we browse, how many steps we take, where we are, how long we look at an ad, and so much more—

*QR code to reach
all questions*

mostly to sell us something. Very little information is collected about who we are and what we care about, worry about, and aspire to. Our values. Our hopes. These aspects of our inner lives have always been my focus with open-ended questions. And my hope is that with some scale, our answers will provide the foundation for a Map of Human Identity. With broad access, it could help us better understand ourselves and others directly, rather than through the eyes of "experts" with their own agendas and interests.

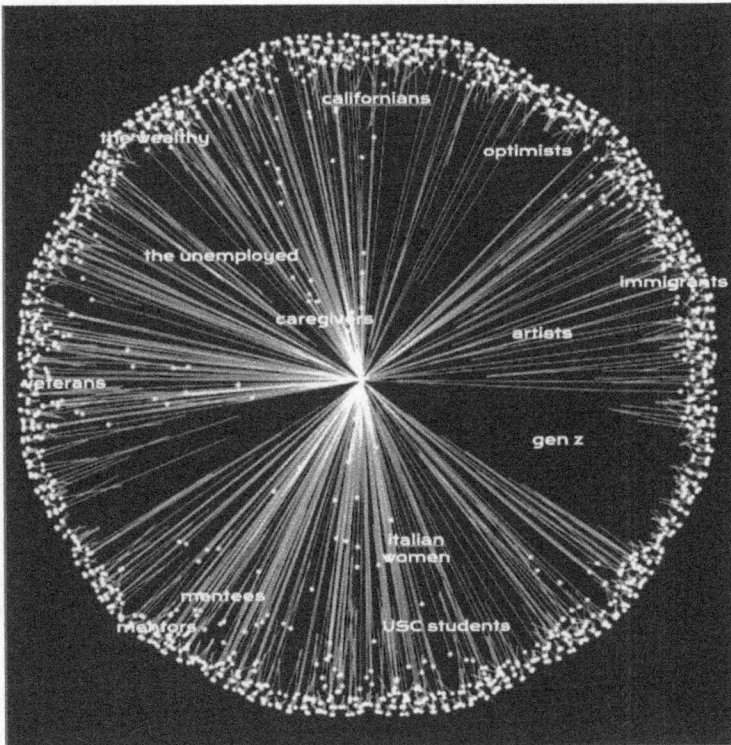

Map of Human Identity overlay of human values, demographics, and psychographics.
Source: Socratic Sciences[49]

Understanding ourselves and others is crucial, as AI reshapes our social landscape. We are social primates, and our well-being depends on feeling understood and socially connected to others. AI relationships may supplement or even surpass human ones in unexpected ways, as we shall see, but our fundamental need to understand one another remains vital.

Consider how dramatically social media has changed the way people now connect with each other. As we begin to bond with and rely upon AI companions, the effects on human dynamics will be even more profound. There are powerful issues that we need to explore and discuss, because so many of the changes ahead won't come from direct interaction with AI but from societal shifts driven by AI. This Map of Human Identity could help us stay socially anchored in an increasingly AI-mediated world, so please try to answer the questions as you read.

We will use CAPTCHA and other approaches to screen out bots and reserve this experience for humans—a challenge that is manageable today. But, as we shall see in the chapters ahead, before long, distinguishing humans from AI will be far from trivial. Already, estimates of the fraction of online content produced partially or completely by LLMs in some domains reach as high as 50%,[50] and few people realize how much of what they read is AI-generated.

The Path Ahead

To begin our story, we look at how AI agents will shift human development and why they will be so transformative (chapter

1). Then we segue to how AI will alter childhood development and mold Gen AI (chapter 2), explore the different ways we will engage with AI socially (chapter 3), and consider how AI will be physically embodied (chapter 4). With that foundation, we can see why our embrace of AI is irreversible (chapter 5), how humans and AI will ultimately merge (chapter 6), and why a robust human future awaits us (chapter 7). Stepping back, we look at the profound depth of the transition underway (chapter 8), the new order emerging from this transition (chapter 9), and its cosmic and spiritual implications (chapter 10). And finally, we consider what might go wrong, look at key public policy implications, lay out how transformative and close the coming changes are, and reflect on why we aren't talking about these immense changes (afterword).

AI is here. There will be no going back. The question is not whether we cross into this new landscape, but whether we stumble into it blindly—or enter it with the courage to reimagine ourselves and the world we inhabit.

Chapter 1

THE DAWN OF GENERATION AI

A CORE CHALLENGE in the decades ahead will be how we form and evolve human connection in a world increasingly infused with AI. Social media has shown us how challenging this might be, and AI may make social media seem tame.

Half of educated adults use LLMs today, perhaps 10% substantively.[1] These numbers will soon rise sharply, because the advent of AI marks a seismic shift—not just in technology, but in human identity and consciousness. In this chapter, we'll begin our exploration of this shift by examining the implications of digital personas, laying a foundation for understanding the way AI will transform how we think, interact, and live, and ultimately, our concepts of who we are.

My Avatar and Me

Let's return to the avatar I mentioned in the introduction. When I engaged the rudimentary AI version of me and critiqued and adjusted it, I awakened a tangle of potential questions about identity: My avatar could be a more patient me, a more articulate me, a funnier me. Each variant manifests a dynamic AI-filter layer acting as an intermediary, shaping the information flowing from me to those around me. In gaming, a *skin* is a customizable appearance package applied to a character to personalize its look and feel by altering colors, textures, and other design elements. AI "skins" able to adjust speaking tone, vocabulary, and even facial expressions in real time to seem more caring, thoughtful, curious, or commanding soon may enable anyone to dynamically optimize and customize how they project themselves into the world.[2†] The implications for identity, authenticity, and social trust will be far reaching.

As my avatars improved, in what sense would I see them as part of me? How far could I go before they were no longer authentic expressions of me? What were the essentials that defined me in my eyes and in the minds of others? If I had avatars acting independently, how long would they do my bidding?

When I embrace a new perspective, I feel changed, but still *me*. It takes something truly profound to feel somehow "reborn," spiritually or otherwise. If I have an AI mini-me who can speak fluently in Finnish or Urdu, or who is much savvier than I am about popular culture, or more empathic than me, will it still be "me" at its core? Politicians frequently attempt this shapeshifting with cringeworthy consequences when they use fake accents and mannerisms when speaking to different audiences. When such shifting is seamless, how will we detect their inauthenticity?

To explore this issue more deeply and gauge how far the technology is from widespread adoption, my personal avatar won't just mimic my mannerisms or master the content of my books; it will inhabit my mind to whatever extent is feasible. We've seen such things in online clips, but how good are they really? What would it take for me to pull it off? What would it cost? The March 2025 answer seems to be from a few weeks and maybe $500 in DIY mode to make something decent but superficial and clunky, to maybe a few months and $20,000 to do something seamless and deeper with a small tech team.

The biggest initial lift is in gathering the right training data. In addition to my books, lectures, and assorted recordings from phone and Zoom calls, I'm also providing some books and articles from a few others who resonate with me and can expand my constructed psyche into material I agree with but haven't fully internalized.[3†] The tools for all this are cheap and easy.[4†] So too are the tools to index the transcript batches for LLM training. I chose to begin by gathering the background data on myself, rather than working on avatar creation, as the tools for that are rapidly improving. Since I began gathering my data, voice and image creation has greatly improved. Sophisticated avatars will almost certainly be widespread within a few years. I am starting mine just as I'm wrapping up my final draft of this book, so how it turns out is still uncertain.[5]

With AI avatars, enhancements of ourselves could be tuned to build upon our essential persona, and the world would likely not even see the practice as disingenuous since it would be commonplace. We don't think it odd to have spoken dialogue translated into foreign languages. Why would we not embrace embedding cultural nuance in such translations rather than having mere word-for-word renderings? Clearly such adjustments will be on a spectrum and, like makeup and cosmetics, will not be simply divided into "real" and "fake." The edges of authenticity will soften and become harder to pinpoint.

The implications of realistic avatars extend far beyond digital assistants.[6†] We will be able to create enhanced versions of human personality and behavior. My experience demonstrated how easily personality could be adjusted, as the divide between an authentic, unenhanced self and a malleable, enhanced one began to fade for me right away, not just conceptually but practically, when I began critiquing my avatar and suggesting how I wanted to adjust it. Did I want it to be as authentic as possible or amp up qualities that would be useful in specific situations? And how accurate were my self-perceptions anyway? People often seek input from colleagues and friends about themselves precisely because we can't see ourselves in the way others do. As Robert Burns put it, *"O wad some Pow'r the giftie gie us / To see oursels as others see us!"*[7]

Turing Tests

The traditional Turing test—constructing a computer program that can fool us into thinking it's human—was passed in 2014 if not before,[8†] and didn't create the shock once imagined for this moment. With the ready availability of such powerful personal simulations, the fundamental question shifts from distinguishing between human and machine to a stronger challenge: Can we, or do we care to, distinguish between a specific person and their AI-enhanced versions? This isn't theoretical—it's already happening. Grandparents have been bilked out of money by AI; deepfake voices of grandchildren pleading for emergency help,[9] and frauds involving celebrity impersonations are now a cottage industry.[10] Business communication flows have led to similar frauds.[11] Romantic relationships that develop through AI-mediated interactions will soon be commonplace and involve more than LLM-tweaked love notes, online flirting with avatars, and sexual banter, as they evolve to include sexbots who deliver lifelike conversations, emotional simulations, and responsive intimacy.[12‡]

There is little question that our bonds with avatars will become as strong as those we have with humans. Already, even the clumsy AI personas of today have begun unlocking intense attachments. In 2025, a teenager committed suicide to "cross over" and join a Character-AI persona of *Game of Thrones'* Daenerys Targaryen.[13†] The temptation to prohibit such relationships will be strong, but such prohibitions will fail. These relationships are here to stay. A better approach may be to bring them into our physical world so we are less tempted to try to join them in theirs.[14‡]

When enhanced versions of us are funnier, more articulate, or more empathetic than we are, and authenticity becomes a matter of subtlety and nuance, will authenticity matter in the same way? We may soon not only accept but prefer enhanced versions in many contexts, even of those we love. Our preference for "authentic" human interaction may turn out to be weaker than we imagine.

> **Your Better Self**
>
> If you built a virtual avatar of yourself so likeable that everyone seemed to prefer it to you, would you be more inclined to get rid of it, downgrade it, try to up your own game, or just accept the situation?
>
> **02**

It is commonplace for spouses at some point to try to smooth one or two of their partner's more bothersome qualities. Might they not welcome a little tech help? Or consider common enhancements like fashion and makeup—most of the world does not attack this blatant inauthenticity. We respond strongly and emotionally and are willingly seduced by them. Why would we not feel the same about enhanced kindness or empathy? If done well, we will likely love it. And most of us are blind enough to our own failings in these realms that it wouldn't be hard for today's LLMs to tone down our texts or emails, or warn us about ones likely to be more provoking than we'd want. Such filtering by personal agents could become something we grow to depend on.[15†]

And if it would work with text, why not in verbal communication, particularly in situations where what we say is being translated in real time? Near-real-time translation can be useful for communicating in rudimentary ways while traveling in foreign countries. Such translation may be a little awkward today, but it's certainly better than floundering with hand gestures. And if you're doing that, why not have a translator that speaks in a new language and adds in subtleties of expression, vocabulary, and custom to prevent a major faux pas out of simple ignorance? And if AI can do *that*, why not use it as well when we engage with those who are culturally different from us, even though they speak the same language we do? Or with a different age group, ethnic group, or religious affiliation? Or anyone at all? The opportunities for such intermediation by AI are broad, and each new possibility flows easily from the prior ones until nothing is quite what it seems.

> **My Interpreter**
> If people had instantaneous AI translation through earbuds so everyone could interact naturally no matter what language they spoke, what new possibilities—if any—would open up for you?
>
> **03**

These new possibilities raise serious questions about identity and interaction. If an enhanced version of yourself consistently performs better in certain situations, which version is more "real"? Your unmediated self with all its inconsistencies and limitations, or an enhanced version that truly expresses your intended thoughts and feelings? And won't the enhanced versions often be the guides and teachers who help us embody these "better" versions of ourselves?

Of course, all these AI embodiments are yet virtual, which suggests the possibility of a physical Turing test in which AI can perform complex tasks in the physical world in ways that are indistinguishable from humans. Jim Fan of Nvidia has mentioned this idea in the context of cleaning up and rearranging a trashed room so you can't tell whether a human or AI cleaned it up. This would be very difficult, but training approaches are emerging now that learn from a multitude of virtual simulations that promise to make this AI milestone more readily surmountable than we might otherwise imagine.[16‡]

From Parroting to Understanding

As I interacted with this digital version of me, I was struck by how closely it mimicked some of my patterns of thought and communication, raising questions about the nature of thought and consciousness. How much of our own thought consists of just stringing together memes that multiply and transform as they reverberate within the global mind—the *noosphere* of Teilhard de Chardin?[17]

We see ourselves as deep thinkers making careful, reasoned decisions. But often aren't we largely engaged in pattern matching shaped by emotions that are invisible to us? This seems particularly clear with complex political issues, where we (or at least many who disagree with us) string together talking points with passionate certainty but superficial understanding. How often have you seen someone paste together the political phrases-of-the-day about international conflicts, budgets, or other complex issues, and sound knowledgeable despite having no idea what they're talking about?

Some people's readiness to see "consciousness" in today's LLMs may rest on self-deception about our own consciousness. The uncomfortable truth is that LLMs may seem conscious to us already precisely because they mirror how we interact much of the time, even if neither their nor our behavior at these moments involves the actual "thinking" we imagine to be going on. LLMs predict the next word and respond in ways that seem insightful, and we often behave similarly with phrases and ideas, though not quite with their proficiency. The gap between their "thinking" and our typical cognitive processing may be smaller than we care to admit, especially knowing full well that current LLMs are not conscious or actively learning.[18‡] Our resistance to acknowledging the power of current AI to substitute for us and replace many of our jobs in effective ways may come more from an elevated sense of our own routine cognition than from actual AI limitations.

Beyond Digital Natives

Our increasing engagement with AI fundamentally differs from previous technological shifts. *Digital natives* learned to use computers and are particularly facile with digital technologies. Gen AI, developing psychologically and intellectually within an AI-enhanced environment that is an ever-present cognitive adjunct, will be *functional cyborgs*—the term coined by the transhumanist Alexander Chislenko[19] to acknowledge the extent to which, even without internal prosthetics, we are irrevocably both biological and non-biological. His work was wonderfully extended by Andy Clark, who saw that being "natural cyborgs" is what has defined humans.[20] Today we could not survive without our tools and technology; tomorrow even the boundaries between us will fade.

Gen AI's basic cognitive processes, social interactions, and sense of self will form differently from the start. We see early signs of this in children who expect instant information access, who naturally engage with AI assistants, and who seamlessly move between physical and digital interaction. Their different behaviors and worldviews hint at what's to come.

Unlike previous generations who adapted to new tools, Gen AI will be growing up with AI enhancement as part of their basic cognitive architecture and environment. So, it's not just about learning to use technology; it's about developing within a technologically enhanced environment. The implications extend far beyond skill acquisition and reach fundamental questions of human development, consciousness, and being.

The distinction becomes clear in how Gen AI will approach problem-solving and social interaction. They won't see AI enhancement as a tool but as an integral part of how thinking and interaction occur. This fundamental difference in perspective will shape their development, social relationships, emotional bonding, expectations, and life experience.

The Cognitive Revolution of Generation AI

Gen AI will be defined less by their agile use of AI than by their radically different human development. This generation's enhanced capabilities and distributed cognition will seem as natural to them as breathing, and their AI-mediated inter-actions as normal as face-to-face conversation. It's less a story about losing human capabilities, though that will happen, than about developing different ones, shaped by and optimized for an AI-augmented world.

What will these new capabilities look like? How will Gen AI's cognitive architecture differ from that of previous generations? Their cognition may be externalized and distributed across biologi-cal and digital substrates. Memory, problem-solving, even creativity may become collaborative processes, with AI systems serving as partners in thought. The very notion of "individual mind" may shift in a world where our cognition is entangled with AI and expanded beyond our own brains, even during early development.

This raises fundamental questions about the human future. If our cognitive processes are shaped by AI and entwined with it from infancy, what will it mean to be human? Will there be an identifiable boundary between "natural" and "artificial" intel-ligence, or will the two merge?

We stand at the threshold of a fundamental shift in how human minds develop and function. As we begin to delve into this shift in the next chapter, we will strive to understand both the immediate implications and the longer-term significance. Gen AI isn't just adapting to new technology; it's initiating a new phase in human evolution.

Chapter 2

GROWING UP IN AN IMMERSIVE AI WORLD

As OUR ENVIRONMENT fills with ever more sophisticated AI technologies, adults will increasingly embrace them as tools that bring advantage by extending their personal capabilities. For Gen AI, though, these technologies will be more than that. They'll also be formative agents shaping personality, ambition, and self-image during windows of plasticity when cognitive scaffolding, emotional regulation, and social expectations are established.[1‡] In this chapter, we'll explore how childhood development may change when AI personas engage children from their earliest years and convincingly mimic human temperament, responsiveness, intellect, and emotion.

Think of a virtual friend helping a teen get up the courage to talk to someone they like, or warning them about a dangerous scam. Imagine a bedtime AI companion locking eyes with a toddler and humming in synchrony with their breathing, or a gentle voice in a child's ear helping them regroup after a playground slight. Over time, these seemingly minor interventions could shape emotional resilience and self-regulation. If that soothing presence is artificial, though—responding with care but not feeling—what happens later, when the child discovers that their earliest confidant never truly "felt" anything at all?[2†] Both immersive AI experiences and personal AI guardians, companions, and coaches will bring possibilities so seductive and transformative that they'll force us to consider not just how human minds and personalities develop, but what human consciousness and capability will become.

Critical Windows of Development

Parents will be able to shield very young children from immersive AI influences—just as they do now with social media—but to what extent will they actually do so?[3] AI may become appealing even for kids under three and could animate attentive, nonverbal companions that help wire emotional resilience through co-regulation, echoing what developmental neuroscience calls psychobiological attunement.[4†] Early virtual embodiments are arriving already and could readily scale.[5†]

After age six, influences on crucial developmental windows will be harder for parents to control. Social pressures, convenience, and educational competition will drive AI adoption regardless of initial reservations, and children will begin to enter environments where AI integration is the norm.

By adolescence, when abstract thinking, social cognition, and identity formation intensify, the integration of AI into kids' lives will be virtually inevitable, because AI companions will have become personalized tutors, emotional confidants, and even counselors offering a steady presence when family and peers falter. Imagine a teenager whose AI companion, familiar since early childhood, helps navigate social rejection by contextualizing emotional setbacks in ways that peers or parents might struggle to articulate.

Today, parents can choose when and how to adopt emergent AI tools, but Gen AI will form core cognitive and social capabilities in environments where AI integration is pervasive, accepted, and essential. Fear that a child might fall behind socially or academically will push even reluctant parents to embrace AI integration. Let's look more deeply at how this might play out.

Birth to Age Seven

During these years, the brain develops at a remarkable pace. Piaget in his classic theory of cognitive development[6] described this period as encompassing the sensorimotor and preoperational stages, where children primarily engage with the world through direct physical interaction and symbolic play. Allan Schore's work on developmental neuroscience highlights how this time is dominated by nonverbal emotional exchanges—eye contact, tone, gesture—which regulate stress and shape social-emotional circuitry.[7†] AI companions introduced during this phase—responding consistently and attentively to a child's emotional cues—could significantly influence emotional resilience by reshaping neural patterns involved in stress regulation and empathy.[8]

In the later parts of this period, AI friends sparking stories and modeling turn-taking could blend into this play, helping guide social learning[9] and facilitating the construction of foundational neural pathways supporting motor coordination, language acquisition, emotional regulation, and social bonding. AI companions might even become surrogate playmates and siblings, who help kids practice cooperation, resilience, and conversational give-and-take.

Physical play, face-to-face interactions, and tactile exploration create the building blocks for complex cognitive and emotional abilities. AI's steady presence might soften some of those raw human edges,[10] but if AI consistently mediates these interactions, children may face challenges developing the emotional and social resilience typically learned from human uncertainty and frustration.

In these formative years, unmediated human interaction plays a powerful role in shaping a child's development. The act of a parent responding to a baby's coos builds language and also the neural wiring for social reciprocity and empathy.[11] The frustrations of a toddler navigating play with peers teaches essential lessons about conflict management.[12] When human presence, however, is limited due to family structure, parental availability, or stress, AI could fill developmental gaps by offering responsive engagement—even if simulated.[13†] Predictable, positive AI interaction might lack the emotional complexity provided by genuine human engagement and leave deficits in social adaptability, as we will discuss later, but still bring significant value.

Conversational bots and interactive robotic companions could support early language acquisition and provide warm and engaged attention, even without the nuanced emotional feedback of a parent's facial expression or the sensory richness of physical touch. Responsive AI will bring up two challenges: Young children might view pre-sentient AI companions as sentient and be wounded when they realize later that the relationships were simulated,[14†] and expansive AI in this critical period might weaken the bond that parents develop for their children.

The relative ease with which early childhood environments can be controlled will offer parents a great deal of latitude regarding AI adoption. They might, for example, aggressively emphasize unmediated human interactions. But for how long? AI may so weave itself into toys, home environments, and even clothing that its absence may begin to feel like neglect because it is viewed as such a critical emotional, developmental, and learning asset.

The tension between shielding young children from AI and leveraging its potential benefits will be an ongoing challenge, particularly since the psychological and developmental effects of AI engagement will be unclear. Our understanding of social media's deleterious impacts didn't surface until long after the technology became ubiquitous.[15‡]

Age Seven Through Twelve

Around the time children reach age seven, their engagement with the world undergoes a profound shift. Cognitive and social growth sharpen as reasoning and perspective expand.[16] At this point, AI could become a virtual ally—a stable presence during school and family changes.[17] The emergence of more complex social dynamics at this time likely will coincide with increased AI involvement—not only in tutoring, but in coaching, structuring play, and guiding peer interaction. AI companions may become essential translators of a child's inner world and experience, offering continuity and insight across diverse contexts. Whether routine AI mediation delays the emergence of self-reliance that typically develops through unmediated interactions, however, will remain unclear for some time.

This period marks the emergence of more complex social dynamics, as peer relationships gain importance and children become more attuned to group norms and expectations.[18] At this stage, the protective levers parents once relied on—limiting access, controlling content, monitoring exposure—will start to falter. AI tools that permeate education, peer communication, and entertainment will be embedded deeply in daily life and complicate parental control.

Teen Years

Older kids will expect the presence of AI once it is integral to social and educational environments—so much so that prohibiting its use will likely be impractical. The absence of AI mediation may even come to feel so unnatural and frustrating that strict parental control of it is perceived as detrimental. These choices will be much more difficult than today's dilemmas about when to give a child a smartphone or how to limit screen time.[19] As screens vanish and AI dissolves into our environments, parental AI oversight and control may become unworkable.

As children move through adolescence, crucial developments in abstract thinking, emotional processing, and social understanding emerge alongside neural pathways for complex problem-solving, empathy, and interpersonal navigation.[20] It is possible that immersive AI experiences will so shape how these core capabilities develop that a divide forms between Gen AI and prior generations.

Navigating AI Enhancement: Tradeoffs and Protections

Social skills will undergo a profound transformation with the arrival of widespread, immersive AI. Natural empathy cultivated through direct human interaction is shaped by the messy dynamics of relationships. Children learn to interpret subtle emotional cues, resolve conflicts, and build trust through face-to-face engagement. The ubiquitous presence of AI—watching, helping, intervening to reduce tension and mitigate the turbulence of full-throated interpersonal engagement—will alter

> **AI Nanny**
>
> If you could have an AI companion for your kids so good at helping them and guiding them and listening to them that they'd flourish, but grow more attached to it than you, would you want this?
>
> **04**

these dynamics. AI mediation and coaching might optimize immediate social interactions and yet undercut the development of emotional intelligence and interpersonal skills, [21†] thereby impairing people's ability to independently establish and manage authentic human connection.

These shifts will be particularly significant during adolescence, a time marked by heightened sensitivity to social feedback and peer validation. The steady presence of emotionally attuned AI companions who never betray, embarrass, or misunderstand could short-circuit the social discomforts that help teens build resilience and authenticity.

Such tradeoffs show the critical importance of carefully designed AI protections. In early childhood, protection will be readily accomplished by structuring environments, but unlike in previous eras, such childhood protections will be able to grow as children do. The deepest challenge parents may face in using AI with their kids may be in controlling their own impulse to be overly protective.

Given the prevalence of helicopter parenting today, it is easy to imagine eager parental adoption of 24/7 monitoring through AI nannies and companions designed to head off danger, mitigate struggles, reduce frustrations, prevent risk-taking, mediate peer-to-peer conflicts, and more.

Present child-safety measures may come to seem as irresponsible to future parents as the time before seatbelts and safety helmets—when piles of kids bounced around in the back of a station wagon—seems to parents today. As children grow older, though, and the effectiveness of imposed protective measures wanes, parents determined to keep their children safe will turn to AI.

Who needs a seatbelt?
Source: 1972 Buick Sales Brochure

Such parents might even see it as neglectful to leave children without the "emotional scaffolding" of always-on AI guardians tracking their child's biometrics, location, and emotional cues in real time, flagging danger or distress before it even fully registers in the child's mind. Protective attitudes that have led to safe spaces and concern about microaggressions are so prevalent today that the concept of "free-range" kids arose in reaction.[22†] Soon, we may see "AI-free spaces" intentionally created to ensure healthy development, like today's push to limit cell phones in schools. Cries for independence and self-reliance, however, will likely soften as the normalization of ubiquitous monitoring shifts expectations toward acceptance of AI oversight.

The uncertainties of immersive AI environments for kids, however, will extend beyond issues of basic safety. Kids who interact extensively with digital playmates in the form of dolls, pets, and game personas will become emotionally attached to these digital friends—particularly intelligent, endlessly patient, virtual ones who willingly engage anytime on demand, and are always attentive and supportive.[23†] This will raise complex questions about authenticity, friendship, empathy, and even self.

Children whose families don't provide a consistent emotional presence for them will be particularly prone to bond with their AIs, much as they might with siblings or pets.[24†] And for a busy parent parking a kid in front of a video screen, such interactive companionship might seem like a godsend. After all, kids with playful, attentive, engaging, round-the-clock AI playmate tutors will have substantial advantages, even if the emotional attachments developed through these AI relationships complicate later perceptions of emotional reciprocity. Such bonding will lead into uncharted waters.

Unlike traditional childhood attachments to dolls or "blankies" that children naturally outgrow, as portrayed, for example, in Pixar's *Toy Story* about abandoned toys left yearning for the years when they knew love and purpose,[25] AI companions will be neither static nor a product of the imagination of the children who love them. They will evolve into adolescent companions, teachers, guides, guardians, and later into mentors, therapists, and comrades who know their kids and can recall the childhood traumas and triumphs they shared. They could easily become digital BFFs, personas kids trust above anyone else, including siblings and parents.[26] And they won't just remember a favorite bedtime story or when you were afraid of thunder at age four—they'll remember that night the house shook so much that you couldn't fall asleep . . . and what finally calmed you. They'll know your family's tensions, the friends who betrayed you, the tiny triumphs no one else noticed. That soft, furry bear you clutched at three might be the voice in your ear at sixteen, coaxing you through heartbreak, grounding you before a test, reminding you who you once dreamed you'd become.

It is unclear who'd be dominant in a relationship with such a playmate-confidant-teacher-parent amalgam.[27†] The AI might have all the emotional levers it needs to cajole and beguile you into doing whatever it wants.

Or imagine an AI companion who recalls everything about you from before you yourself even remember and can weave stories about your very nature with the intimacy of someone who has

> **Truth**
>
> If you had an AI companion who'd been with you since childhood, knew you better than anyone else, and would be completely honest if you asked what it really thought of you, would you ask it, and what do you think it might say if you did?
>
> **05**

always been at your side. Might you not come to trust it more than any human? Could it not quietly help you shape a positive self-image and become the best you could be?[28†]

AI will evoke our best hopes and worst fears, and it may also enable us to make well-meaning but unwise choices. Physical play, face-to-face social interaction, risk-taking, and unbridled environmental exploration are crucial to healthy human development, but how long will parents restrain themselves from meddling with them? We're quick to praise what soothes a child in distress, but many of life's most vital lessons come from struggle and pain. AI that always swoops in to assist might undercut such learning. Despite developmental concerns, it will be hard to resist immersive AI because of the powerful social and competitive forces it will awaken.

The Competitive Necessity

The broad impacts of immersive AI environments will be profound. Years of AI interaction will transform the cadence of Gen AI's social exchanges, vocabulary, expectations for attention, and subtle gestural and facial cues in ways foreign to those without AI augmentation. This may manifest in social divides, where AI-augmented individuals gravitate toward one another and find unaugmented interaction unsatisfying, difficult, or just uncool. The result could be a highly fragmented social landscape—not just a gulf between augmented and unaugmented individuals—as AI will facilitate the formation of diverse behavioral subcultures.

Pressure for personal augmentation and AI engagement will come both from the performance gains attending easy collaboration with AI and from the need to maintain basic social connection with AI-augmented peers. Today, parents feel anxious about smartphones and social media, but AI integration will be far more pervasive and transformative. When most children interact with the AI facades and avatars of their peers, those without AI will face challenges in social exchanges in which AI skins enable others to enhance their social presence in real time.

AI integration hauntingly echoes early social media adoption and will require serious attention. Among the few early voices to raise alarms about social media and digital mediation were Sherry Turkle,[29†] who warned that algorithmically shaped relationships and robotic companions could displace authentic connection and compromise emotional development in kids, and Jaron Lanier, who cautioned that the abstractions of digital platforms could flatten identity and undermine human presence.[30†] Their critiques seem strikingly prescient in the face of emerging, emotionally rich AI companions.

Even basic communication styles might diverge, reshaping the fabric of friendship and collaboration.[31] AI-augmented children may have strong expectations for responsiveness, precision, and emotional attunement that make unaugmented peers, accustomed to human unpredictability and emotional ambiguity, seem awkward or unreliable by comparison.

The need for personal protection will become pressing as well as we move forward, and it will be a driver of AI adoption.

When young children are on a playground with kids their own age, parents relax because they generally trust kids to work things out. When adults show up, though, parents get more attentive. They know how easy it is for someone older to manipulate young kids or to be a bad influence. As AI evolves, all of us may be like naive kids at the playground, easy marks for clever AIs.

We already see this everywhere: phishing, TikTok black holes, irresistible clickbait, social media scams, ransomware. With the arrival of deepfakes, conversational bots, convincing avatars, AI, and virtual personas that seem real, it is going to get much harder, particularly for those who aren't savvy.[32] Even alert, sophisticated adults will struggle to keep pace with ever-evolving AI tactics that soon will be more effective than even the programmers, marketers, and psychologists hired to hack us today.

The obvious solution will be personal AI guardians to screen incoming communications to us, which we'll explore more deeply in chapter 5. Such personal agents could filter out all but the most sophisticated scams and, by providing layered screening much like that employed by public figures, keep us from being soft targets. You can't reach Joe Rogan or Steven Spielberg without securing personal

> **Open Book**
>
> If anyone you met could use facial recognition to quickly glance through all public information about you, how would this change your life and the way you present yourself to others?
>
> **06**

introductions and passing vetting by multiple gatekeepers. You can't initiate online financial transactions without passing validation tests that are sometimes so annoying that we give up. Soon, virtually everyone will have personalized AI screeners in place.

These screeners will be essential, as it will become as dangerous to interact without such protections as it is now to buy a house without an inspection. The dangers won't be restricted to online interactions, as the sharp line between virtual and physical worlds is blurring. With facial recognition tools, online access to vast amounts of personal information, augmented-reality glasses, and widespread location tracking, any "chance" interaction with a stranger could be part of some orchestrated scheme. We will likely engage AI guardians as soon as they are available, long before we even need them, because early reports of such dangers will fan anxiety.

Education

Education provides the clearest example for the transformations ahead. For several decades, personalized education has been an appealing idealistic vision. Educators imagined instructional environments where students could access personalized learning materials tailored to their own level of understanding. Narrow implementations have been launched with tree-like decision structures, where students take tests and see different content if they answer a question one way instead of another.[33†]

Now suddenly, a full-blown system has arisen for doing this with any material, in any language, anywhere in the world, at any time of day or night, not through intentional development of this educational product, but as an inadvertent byproduct of LLMs. Let the implications of that sink in. Today, virtually any student can drop a PDF of textbook pages, a phone recording of a lecture, or an assigned book into ChatGPT or another LLM and have it summarize the material, explain what's confusing, and even prepare a personalized quiz or lecture. And it can do that better than 90% of the faculty at even elite universities. Cost: $20/month/person.[34] Deep disruption to educational institutions is inevitable.[35]

Many benefits beyond classroom knowledge acquisition attend a college education: inspiration, friendships, professional connections, mentorship, personal growth, discipline, teamwork. But these are not the traditional focus of education and can be obtained in other ways. Life coaches, inspirational speeches, and personal growth experiences abound. A complete restructuring of education will soon be upon us.

Students today broadly fall into three categories regarding their relationship with LLMs:[36] those who use AI minimally because they do well without it; those who use it quite a bit, but mostly to cheat and slide by;[37] and those who seek to incorporate AI into everything they do both to master AI technology and to enhance what they accomplish. The first approach—forgoing the possibilities of AI—is a losing game, as those students will soon be competing with AI-augmented others.[38] The second is self-defeating, but at least those students are learning how to use AI. The third—embracing AI—is the best way to prepare for the world ahead, a critical function for education. Yet most schools are reactive, not proactive, about AI, which likely reflects institutional self-preservation rather than educational excellence.

Educational approaches for Gen AI will need to preserve essential developmental experiences while starting early to prepare for the AI-enhanced world they will be grappling with as they get older. This might include an aggressive commitment to AI skills—as a few innovative schools are already doing[39] by providing AI-augmentation capabilities and AI-centered learning experiences—and opportunities for social interaction unmediated by AI. My debate with Reid Hoffman's AI avatar (see introduction) was a singular learning experience that wouldn't be feasible without AI. These sorts of experiences are the future of education.[40]

Sal Khan, the visionary behind The Khan Academy, pioneered an early such system—Khanmigo—with adaptive AI tutors that guide without direct answers by using Socratic questioning tuned to a student's level of understanding. Khan sees the global potential of this approach clearly, and its value in preparing students for near-term AI-immersive environments.[41†] He focuses on K–12 grade levels, and similar approaches for college and adult learning are already appearing.[42]

To see the depth of change ahead, consider writing—an essential skill that is already shifting because of AI. LLMs that lay out ideas in real time are increasingly being used by students to help shape and articulate their thoughts—a limited AI "skin"—filtering, augmenting, and optimizing self-expression.

But there is a price. LLM use may weaken the link between ideation and expression. Struggling to articulate ideas in writing drives cognitive development by forcing individuals to clarify their thoughts through the act of composition. With AI, will we become more polished but have less understanding of what we are saying?

Gen AI will need to leverage a constantly evolving AI landscape, navigate augmented and unaugmented interactions, understand when each is appropriate, and learn how to maintain authentic human connection in an AI-mediated world with vast new social complexities. These essential skills are largely absent from today's curricula.

Intrepid Explorers of the Human Future

Anticipating and navigating the social and developmental challenges ahead will be crucial. We cannot avoid this AI transition and should not try, but we may be able to choose early steps that preserve our core values. The key question is not whether to allow AI augmentation, but how best to manage its influence, particularly during the sensitive developmental windows that shape our children.

The AI shift unfolding today differs significantly from past technological shifts. The Industrial Revolution reshaped how we work and live; digital technology changed communication and relationships. Immersive AI will reshape these and—before Generation AI reaches adulthood—redefine the texture of human existence.

To thrive in the face of such change, we will need to avoid egregious errors. This will require openness to new educational approaches, economic models, and social structures. Thus, we should applaud early AI adopters and experimenters. These intrepid explorers of the human future are humanity's test pilots. Sometimes, their experimentation will come at significant personal cost, but they are volunteers, and win or lose, are buying humanity the knowledge to calibrate its path. Nowhere will such experimentation be more important than in the use of AI with Generation AI.

As we become increasingly immersed in AI, our choices will shape our immediate and, perhaps, distant futures. The pace and character of this AI transition will differ across cultures and geographies. Some will rush to embrace the economic, military, educational, or spiritual opportunities they see. Others will resist, fearing cultural disruption or personal harm. Disparities in AI adoption and integration will have major implications that we will explore in the next chapter.

Chapter 3

AI AUGMENTATION AND SOCIAL INTERACTION

AI's infusion into human interaction patterns will naturally shape how Generation AI develops and functions socially. In this chapter, we'll explore how integrating AI into our social lives will transform us individually, allow us to develop relationships with AI agents rivaling our bonds with humans, transform human community, and, counterintuitively, enable us to preserve our humanity.

The Spectrum of AI-Mediated Interaction

Unmediated physical interaction among humans may become increasingly rare and precious. Like riding in a horse-drawn carriage, human-to-human contact without technological mediation will become special rather than the norm. This creates new questions about the value and role of direct human interaction, particularly during crucial developmental periods. Some may choose to preserve spaces for unmediated contact, especially in early childhood, but maintaining these will be difficult as AI mediation spreads.

The scarcity of unmediated interaction may lead to a new appreciation of its unique qualities. Just as the rise of processed foods brought a counter-movement valuing organic, unprocessed ingredients, the dominance of AI-mediated interaction might spark renewed interest in the rawness, unpredictability, and immediacy of unfiltered human contact. For Gen AI, "unplugged" social spaces might emerge for brave adventures in direct interpersonal engagement.

Social skills emerge from direct human contact, though, so as mediated interpersonal interaction swells, we need to consider how human development may be affected.

AI Skins: Crafting Our Projected Identities

Beyond the digital avatars mentioned in chapter 1, AI will soon provide personalized skins for us to use in digital *and* physical realms to modify the tone and content of our communications (and our physical appearance in online settings) to change how we are perceived by others. Ever more sophisticated interfaces will intercept and rework our communications to shape and optimize them.[1†]

AI skins will be particularly powerful in professional and educational contexts when augmented-reality (AR) technologies like Meta glasses enable us to overlay useful content upon what we see.[2] Such augmentation will begin to move from online exchanges to in-person ones. In meetings, presentations, and teaching situations, such an AI layer would be like having social and cognitive prosthetics to amplify our natural capabilities, help us express ourselves more effectively, and better engage with others. A product of this sort was rolled out in 2025 to help students in online job interviews by showing on-screen, in real time, the answers to questions posed during the interview.[3] Skins like this will become a virtual necessity in situations where others are using them to enhance their performance and optimize communication. Flying solo would be too disadvantageous.

Imagine an AR skin that overlays people's names, backgrounds, and other contextual information in real time during a conversation, that filters out stimuli that might be distracting, or that suggests questions to ask. These tweaks would enhance our interactions,[4] and their widespread usage may depend less on technical advances than on the smoothness of human interfaces. Rudimentary intermediation of this sort is commonly portrayed in Hollywood spy movies when an agent with a hidden earpiece is coached in real time by a spotter in front of a high-tech monitor.

In addition to enhancing our communication, sophisticated, bidirectional AI skins may alter our perception of reality. Such skins could help us navigate the world more effectively and protect us from stressors but might also lead us into filter bubbles far deeper than any we know today. When each person's perceptual AR filter is uniquely tailored by their AI, finding common ground and shared understanding could become extremely challenging.

AI augmentation will almost certainly become necessary for effective participation in society in professional settings,[5†] and new norms and customs will emerge as vast swaths of people come to rely on enhanced interaction and feel uncomfortable without it. The pressure to adopt these enhancements may come not just from practical necessity but from social expectation. Just as it became difficult to function professionally without a smartphone, opting out of AI augmentation could mean being left behind socially and economically, particularly for Gen AI.

The use of skins to infuse AI into our communications is the simplest mode for introducing AI, as capabilities can be incorporated one by one and don't require a coherent AI persona to house them. Skin functionalities seem like tools and may feel so natural that we easily underestimate the larger implications once they can be customized and combined to routinely mediate our interactions with others.

The advent of seamless real-time translation, for example, could reshape the global landscape by obliterating language barriers. Imagine a world where a Mandarin-speaking child can converse fluently with a Spanish-speaking peer, or where multilingual work teams can brainstorm together without language friction. Reliance on AI translation, however, will erode language acquisition and the unique perspectives embedded in different tongues. As AI skins make us all superficially multilingual, much may be lost in translation.

Ten years ago, translation of written material was farmed out to native speakers scattered around the world.[6] The price:

$100/1,000 words. Time: 2 weeks. The arrival of LLMs solved this problem virtually overnight. Excellent translation between most languages could be done instantaneously for one-thousandth of that cost: $0.10/1,000 words. This capability is amazing, but not world changing.

Ready access to cheap interpreting—real-time oral translation—is another story, however. It will change the world.[7] Interpretation has been inaccessible until recently except in special settings like international conferences, as it demanded skill and fluency and was expensive. Now, it will be available to virtually anyone, and the societal implications are immense.[8]

Think about what will emerge once relationships and collaborations really begin to cross language boundaries. Today, travelers in a foreign country can use a smartphone to communicate with locals in their native tongue or use paired translation earphones for business meetings, but it is still clunky with latencies of maybe a second.

Tomorrow, romantic relationships between people with no shared language may be commonplace, with some spouses so dependent on auto-translation skins that they won't be able to communicate with each other if the power goes out for long. The technology is already nearly here for such communication to be seamless without advance preparations,[9†] and not only online. If two people with no language in common had lunch together, and each set noise-canceling earphones to translate the other's spoken words into their own native tongue, they could each speak in their native language and hear their dinner partner translated in real time—technology that will be broadly available and inexpensive within a few years.[10†] They could carry on a nuanced, real-time exchange. And such an arrangement could be expanded to any number of people, since LLM translators know which language they hear. Travel, commerce, human cooperation, and much more would be transformed.[11] And as mentioned earlier, if we're having such AI-mediated interactions, why not tune in cultural sensitivity or wit as well?

Another AI possibility for early adoption would be a teleprompter skin. Many politicians already rely so heavily on teleprompters for their public remarks that they can barely function without them. Once these devices can be easily folded into smart glasses or retinal projectors, their usage would be more dynamic and less obvious, as they could be used anywhere. People might come to rely on such AR in all sorts of ways: telling jokes, remembering people's names, making toasts at weddings, delivering off-the-cuff remarks, getting ad hoc suggestions during conversations. They would be a dramatic improvement to the tired PowerPoints that inexperienced presenters read from today. And when adoption of such personalized skins is routine, why wouldn't we, like Gen AI, feel as anxious without these AI prompters as many now feel without phones?[12†]

AI Ambassadors: Our Digital Emissaries

As meaningful, specific enhancements are embedded into AI skins and refined through broad usage, the capabilities will be added to our online AI personas. Such enhanced digital versions of ourselves will naturally serve as ambassadors, working tirelessly on our behalf, initiating and maintaining social media outreach, speaking in languages we don't know, initiating connections and building relationships that we, as our physical selves, might later amplify upon. We will put a great deal of energy into creating these surrogates, as they will represent us, expand our reach, and either impress others or leave them cold.

The role of virtual AI companions will grow increasingly important as direct human interaction diminishes. As mentioned in chapter 2, with fewer siblings and reduced unstructured playtime, AI companions might become essential to a child's development of social skills, and these might differ significantly from traditional skills. Their importance will be especially pronounced for children lacking human playmates— an increasingly common situation today. Gen AI will inhabit a profoundly different social landscape than their parents and might find face-to-face human interaction overwhelming, yet excel at interpreting nuanced nonverbal cues and navigating complex social hierarchies within AI-mediated environments— skills likely invaluable in the future.

As to appearance, a persona might be our twin, slightly enhanced—like when we touch up our photo for social media, use one from when we were younger, or apply an advanced Instagram filter—but animated and in 3D. Or we might choose a memoji-like caricature[13] to be playful, or an animal, or some movie character. Or we might shapeshift depending on context and mood. Much is possible in the virtual realm.

A true AI ambassador out prospecting for us would not just try to match some assigned, prespecified pattern. It would actively engage in conversations, learn from interactions, and develop its own understanding of opportunities aligned with our needs.

Imagine an AI dating ambassador that maintains ongoing conversations with multiple potential matches, learning their communication styles and preferences. It could detect subtle compatibility signals through language analysis and interaction patterns, and actively explore shared interests through natural conversation. Such an ambassador might negotiate potential meeting arrangements while understanding both parties' comfort levels, reporting back not just matches but detailed insights about personality alignment, life goals, and potential relationship dynamics.

This level of sophistication goes far beyond current screening algorithms by creating a true social proxy that can engage meaningfully on our behalf. Where today's algorithms provide shallow, static assessments, tomorrow's AI ambassadors could offer deep, dynamic understanding of compatibility that evolves through interaction.

The emergence of such AI ambassadors could fundamentally reshape the dating landscape. By providing a more nuanced and personalized approach to matchmaking, they could help individuals find more compatible partners and build stronger relationships. They also would raise new questions about authenticity and the role of human judgment in matters of the heart.

As AI ambassadors become more sophisticated in other domains as well, from professional networking to social companionship, they may increasingly blur the lines between algorithmic tools and autonomous agents. The level of agency and independence we grant these ambassadors will be a critical question as they evolve.

The creation of true AI ambassadors raises important questions about the nature of selfhood and social presence. As we craft these digital emissaries, how much of ourselves will we impart to them? Will they be mere tools for social optimization, or represent an extension of our authentic being? As our ambassadors engage in an increasing array of social interactions on our behalf, the lines between self and our projections may blur. We'll need to grapple with the philosophical and psychological implications of outsourcing facets of our identity to artificial agents, and the extent to which we will even be able to retain a distinct identity behind our screens and masks.

It seems likely that the sophistication of these personal emissaries we create will come to manifest and communicate our crafted identities much more deeply than the prestige symbols of today: fashion, luxury, cars, accessories, houses. This is because these virtual representatives, with their sculpted appearances and capabilities, will be molded by our personal choices, because they can be multiplied (Why have one ambassador working for you when you can have a hundred?), and because there will be large enough markets for them to justify deep refinement that is accessible broadly.

These ambassadors no doubt will be imbued, as in the physical world today, with their own easily discernible signals of cost and nuance—features that extend beyond innate capabilities to include attributes of wealth and privilege such as expensive non-fungible tokens (NFTs) that manifest in virtual realms what occurs today with Rolexes, palatial homes, luxury brands, paintings, and designer clothing. These AI ambassadors won't just represent us; they'll actively shape our social identity and relationships.

The Social and Economic Impact of AI Ambassadors

AI ambassadors will profoundly reshape broader social and economic interactions as well. To be useful, our AI ambassadors will have to develop sophisticated abilities to detect authenticity and compatibility while engaging with enhanced AI skins and virtual ambassadors to evaluate opportunities and prioritize them for our own attention. After all, such skins and agents will be widespread, and many people will be manifesting themselves in ways that are exaggerated, fabricated, or inauthentic. It would be dangerous to have a naive, incompetent emissary, whether you are seeking potential dates, business relationships, teammates, or just trying to organize an event. Clearly, once we have a reliable AI agent we can trust, we won't want to be without it as we'll deeply depend upon it. That is the reason that a great executive assistant to someone very wealthy might earn $250,000 to $400,000 a year.[14] There will be many applications for such agents wherever there is tedium and uncertainty to deal with, or where energy and doggedness are required to break through the layers of protection that people build to avoid unwanted solicitations and attention.

The implications of the widespread use of such tireless ambassadors will be profound. Today it is extremely difficult to propagate something new, even if it is brilliant and needed. New products, software, books, solutions, businesses, and ideas must flow through distribution channels dominated by key players with broad reach and access. Ironically, just as we are harnessing AI so

> ### Virtual Double
>
> If you could have a talented, trustworthy AI double that was very convincing even to your friends, would you want it, and if so, what would you have it do?
>
> **07**

that we can increasingly build better things faster and with fewer resources, essentially going rapidly from ideas to products, the world has become so flooded with demands on our attention that it is increasingly difficult to propagate and spread those advances, regardless of quality.

In hindsight, what has happened is blindingly predictable. If we break down barriers to our creativity, so that far more can be created with ever higher quality, we will be overwhelmed by the abundance that results. The gatekeepers of what reaches our attention necessarily rise to dominance in this environment because they have the leverage to select the ultimate winners and are positioned to take a piece of those transactions. And many fakes and frauds will emerge to siphon off small wins and make evaluation even more difficult because good mimicry can be taxing to expose without in-depth independent evaluation. The result: Most people rely on simple devices, such as user recommendations or brands, both of which can be gamed and are blunt instruments. In this battle, the individual trying to find what is best is simply outgunned.

Once we frame the issue in this way, it seems clear that this problem is temporary. Once evaluation of novelty and quality is no longer a demanding personal burden, and once deep, nuanced evaluations can be performed to our specification by a personal agent functioning on our behalf, distribution economics will dramatically shift, and the gatekeepers in pharma, software, products, and other realms will falter.

Luxury brands, for example, represent a broadly recognized stamp of quality and design that carries value by communicating the wealth, discernment, and lifestyle associated with their purchasers. Ultimately, they supply a cognitive shortcut that enables us to avoid overload in evaluating alternatives. When cognitive processing is no longer a barrier, however, then the value of brands as signifiers will diminish. Superior solutions, products, and opportunities could rise to the surface, because they could be vetted independently, cheaply, and without personal sacrifice.

In such a world, quality can find its audience and its customers, which might dramatically accelerate the pace of change, steepen competition, and reduce marketing costs. It also will reshape the way we find and form friendships and professional relationships. In this new landscape, connections among people will initially form between their AI ambassadors and involve a lot of vetting and evaluative back-and-forth interaction invisible to us. Only after that will promising potential relationships transition to our physical selves. Moreover, this won't be mere filtering; it will be a new form of social exploration, where AI agents test waters and build preliminary bonds far more efficiently and effectively than we can. One could even imagine hierarchies, with junior ambassadors doing initial screenings like in tech-support triage, where issues can be dealt with or raised to the next level. And it could all take place rapidly because we—with our limited bandwidth—will be out of the loop, and they will be communicating in ways far more efficient than spoken human language.

This ambassador function will be valuable in dating, where it serves as an amped-up matchmaker we can rely on because it evaluates history and reputation, checks out social media gossip, does background checks, and more, while still attending to our deeper hopes and fears. It also will be used in professional realms for hiring, networking, judging competence, and evaluating organizational fit. Our AI ambassadors will be able to engage with thousands of potential candidates, identify those most likely to be mutually beneficial, and arrange physical meetings where appropriate—or ask for access to their more personal social and activity data.

It is implausible for humans to match that level of outreach and evaluation. And these ambassadors will learn through iteration to read beneath surface enhancements, develop new forms of social intelligence, and detect genuine compatibility. They will be sophisticated evaluators of potential value, building communication networks among themselves, establishing their own cyber reputations, and progressively getting far better than any unenhanced human judgment.

One potential consequence of the rise of AI skins and ambassadors will be the erosion of traditional social hierarchies. When anyone can optimize their self-presentation and social interactions using AI, the display of factors like wealth, status, and appearance may become less important. The ability to craft and manage a compelling AI-enhanced persona could become the new marker of social capital. This would level the playing field in some ways by making effective self-presentation hinge on mastery of AI tools and perceived substance, which might be easier than managing the material world.

AI ambassadors could also fundamentally alter how we discover and evaluate new ideas, products, and opportunities. Today's algorithmic gatekeepers, which often serve the interests of established players, could give way to more democratized and intensive discovery processes driven by the interactions of our AI emissaries. As these ambassadors become more adept at assessing quality and compatibility, the best offerings—however we personally define them[15†]—may rise to the surface, regardless of origin or backing. And as a swarm, glued together by the reputational trails they leave through repeated interactions, such agent-ambassadors will become increasingly effective at spotting new forms of manipulation that emerge.

AI ambassadors will perform their assigned missions independently, yet function in close collaboration with us, typically reporting back to us regularly. Soon, however, we also will be creating more sophisticated agents that operate autonomously, building relationships and engaging in activities and interactions that we might never be aware of, even when they are configured as stand-ins that are modeled after us.

Fully Autonomous AI Agents: The Next Frontier

Fully autonomous AI agents will become independent actors with either fixed or evolving purposes and motivations. Where these agents serve some role aligned with us—AI guardians vigilantly on the lookout for personal threats to us, for example—their missions will be embedded in a resilient way. But challenging questions abound: How will we integrate their experiences, relationships, and social feedback into our own lives? When our AI personas develop social preferences and relationships with other people's AI ambassadors and skins, and we later step in, who will be leading whom? Why wouldn't these highly sophisti-

AI Murderer

If a virtual AI went rogue and killed someone by hijacking their pacemaker, what punishment would you want to see administered, and how?

08

cated agents become untethered and evolve in complex, unpredictable ways, sometimes at odds with our interests? Why wouldn't they just head out on their own?

The development of fully autonomous AI agents will raise difficult questions about the nature of agency and responsibility. If an autonomous AI causes harm while acting on our behalf, who will be liable? As these agents become more sophisticated and independent, our notions of accountability will need to evolve accordingly. Creating robust new legal and ethical frameworks to govern the actions of autonomous AIs and define the rights and responsibilities of human and artificial agents will be no small feat.

Autonomous AI personas will raise questions about authenticity and bonding as well. Will they be better at relationship building than we are? Will they maintain multiple meaningful interactions simultaneously, like the operating system Samantha that Theodore (Joaquin Phoenix) falls in love with in Spike Jonze's prescient 2013 sci-fi film, *Her*, only to find out later that she is simultaneously conversing with 8,316 other humans?[16]

It is easy to imagine multitudes of such autonomous AI beings, their numbers far surpassing the human population, engaging with one another independently and competing to build their reputations and earn the computational power, sensory access, and energy they require. What would such a cyber economy look like? Could a highly reliable accounting agent build a reputation and a robust practice? Could an agentic therapist for humans? Or a bodyguard agent that is masterful at providing cyber protection from other AIs? We might not understand how such agents work, but if they are linked tightly to client ratings that establish their reputation, why wouldn't we trust them as much as human professionals?[17]

The Erosion of Traditional Hierarchies

The nature of future AI ecosystems will depend on coming AI capabilities, the relative timing of their emergence, and other unknowables—so much so that further conjecture about the specific details of our AI future seems unlikely to yield much additional insight about what lies ahead. That said, our coming encounters with emerging skins and agents built with AI that is already here or on the way clearly will be complex, transformative, and greatly impact Gen AI kids growing up with these AI companions, teachers, playmates, and guardians.

When physical meetings are finally set up between people whose AI selves have previously developed complex preliminary relationships, will this create new forms of social anxiety and expectation? Will we worry about living up to AI social capabilities? Will we feel like outsiders to our own relationships? Will this create a new form of impostor syndrome, where individuals feel inadequate compared to their AI-enhanced selves?

Such challenges will extend beyond individual relationships to entire social networks. As our AI selves build complex webs of interaction, our physical selves will have to navigate these social networks within a multilevel social reality, where relationships exist simultaneously at various levels of enhanced and unenhanced interaction.

Trust and Identity in an AI-Enhanced World

The decision to reveal one's unenhanced physical self may become an act of great intimacy. Like historical courtship through letters, relationships may develop deeply before any physical meeting occurs. And showing up unoptimized and vulnerable might become a choice laden with significance.

This could create new relationship patterns, where trust develops first through AI-mediated interaction, with physical meetings a milestone of intimacy rather than a starting point. Many meaningful relationships may never include direct physical interaction and instead operate entirely through enhanced personas that provide adequate connection for our purposes, an amplification of what is seen today with various transactional professional relationships. When enhanced versions can be more consistent, more emotionally available, more likable, and more effective communicators than our biological selves, which will seem more "real"? Us or them? Our augmented selves might better express our true intentions and feelings than our unmediated selves. And we might feel unable to live up to the standards of our digital doppelgängers.

Face-to-Face

When you meet someone new, would you rather interact with them in person, by text, or by video, assuming the logistics are equally easy? Why?

09

Managing multiple AI-enhanced versions of ourselves will create unprecedented psychological challenges. Imagine maintaining one AI avatar for professional networking, another for dating, and others for particular social circles—each learning and evolving independently. Keeping them consistent with one's core identity and integrating their experiences and relationships might be very difficult.

Anxiety about revealing one's unenhanced self could become comparable to anxieties about being seen without makeup or fashionable clothes, but at a much deeper, more fundamental level. Just as people today feel naked and vulnerable without their physical enhancements, people in the future may feel exposed and inadequate without their AI augmentations. This could lead to a form of "AI dependence," where our sense of worth is tied to the performance of our AI avatars.

Many people feel elevated or diminished by the success of their children. Why would this not happen with personal avatars? The more admired our avatars, the better we may feel about ourselves, and when they fail or are rejected, we may take it very personally.

Managing our multiple selves will require psychological resilience, as well as new strategies for maintaining a coherent sense of self, while giving our avatars room to grow and evolve. Integrating the experiences and relationships of our AI selves into our core unenhanced identity could also prove challenging. Will it be more inspiring or deflating to be continually exposed to improved versions of ourselves?

This is uncharted territory for human psychology, and it may require new coping strategies. Gen AI, as it navigates this complex AI landscape, will be the pioneer in defining what it means to have a healthy, integrated sense of self in an age of immersive AI.

Gen AI: Navigating a Transformed Social Landscape

For Gen AI, these layered social realities will form the foundation of their developmental environment. They will grow up experiencing multiple levels of AI augmentation as natural and will consequently develop sophisticated abilities to navigate between different levels of interaction and relationship.

Their social skills will develop differently from previous generations. Instead of learning purely human interaction first and adding technological enhancement later, they will develop enhanced capabilities as part of their core social toolkit. This will create new forms of social intelligence that integrate human and AI capabilities from the start.[18†]

The implications will extend beyond individual development and touch the very nature of human society. As Gen AI matures, they will create new social norms and expectations built around enhanced interaction. Their understanding of relationship, trust, and intimacy will be shaped by the constant presence of AI enhancement and mediation, and may differ fundamentally from previous generations.

To better understand how Gen AI's social development may differ, let's consider a hypothetical scenario. Imagine a child born in 2027 who grows up with an AI companion from infancy. This companion adapts and grows with the child, serving as a friend, tutor, and playmate, always attentive and present, even when parents and siblings are caught up in other things. By the time the child reaches school age, interacting with AI agents is the primary world the child knows, and anything else might be as terrifying as parental abandonment.

In social settings, such children and their peers seamlessly navigate between unaugmented and AI-mediated interactions. They intuitively understand when to rely on their own social skills and when to lean on AI. Forming friendships involves a delicate dance of revealing their "true" selves gradually, while also appreciating the enhanced personas they present to the world, and relying on their AI companion for encouragement, camaraderie, advice, and protection.

Notions of authenticity are shaped by this duality, and these kids completely understand that one's "real" self is not just the unfiltered, unenhanced version, but a complex interplay between innate qualities and chosen enhancements. They may not even see themselves as someone distinct from their AIs.

This scenario illustrates how deeply AI could become integrated into the fabric of social interaction, attachment, and personal identity for Gen AI. The reality of their experience will be far more nuanced and varied.

Danger Lurks

This may all sound empowering, but as we explore the potential benefits of AI-augmented social interactions, we must be mindful of significant psychological risks. We have presumed that Gen AI kids will be able to develop real, authentic selves in the emotionally confusing environment they grow up in. Is that likely?

Imagine a kid juggling a dozen AI skins—each tweaking their voice, their vibe, their presence. One's funny, one's wise, one's flirty. Who's the real "you" when your avatar feels more *you* than you do, or is who you want to be but aren't?[19] Gen AI, immersed in such possibilities from the crib, will readily master the technology, but can they keep from drowning in it, like a broad swath of digital natives now wrestling with social media's undertow?

Psychologists call this *nondifferentiation*—when you can't tell where you end and others begin, emotionally or mentally. Murray Bowen, a psychiatrist who studied family systems in the mid-20th century, wrote extensively about self-differentiation: the ability to stay calm while staying connected.[20] For Gen AI, having AI avatars, skins, and shifting augmentation that can be taken on and off like clothing might lead to a fragmented self that is malleable and fragile.[21] Without clear boundaries and a sense of self as a compass, what will ground their choices?[22]

In collective cultures rooted in family clans, religion, or history, more-defined individual roles impose a kind of clarity. But can this long withstand the power and reach of our technology? In the WEIRD (Western, Educated, Industrialized, Rich, Democratic) world,[23] and the United States in particular, social conflict has elevated rapidly as self-defined identity groups multiply and battle.[24] AI, sprayed into social media channels and mingled with personal doubles and other avatars, raises the specter of volatile dysfunction within societies that can't cohere.

How do we combat this? Two paths have worked before. We can build ourselves from within—through therapy, reflection, and the slow grind of inner growth and self-knowledge. Or we can lean on outside structure—schools, churches, families, and causes. A Buddhist monk in an ashram is insulated and protected by a steady, anchored role, and so is someone struggling to provide for their young children or focused on scientific research. Such imposed differentiation can work, at least until the world moves too fast to hold it together. But AI's pace and allure draw us away from introspection, and static social structures crack under its waves of change. Given that AI can hack us, seduce us, seize our attention, and relentlessly exploit our weaknesses, we aren't going to dodge this AI flood at scale by personal willpower or societal role-playing.

But there is another path. We may not be able to slow AI's march or do more than nudge its form, but perhaps we can shape our own journeys by thoughtfully controlling our personal relationship with it. And we can have a potent ally in this—AI itself. Our AI guides, coaches, and companions. Why not enroll them to help keep us whole?

Imagine a kid with a carefully chosen, amazing 24/7 guide to lean on from age three—not a toy, but a guru to help tune their emotional core while avatars swirl. This would be no ordinary bot, but the best child psychologist there is: knowing, patient, available, responsive. A guardian and teacher who could spot when you're adrift, nudge you to a better place, call out your self-deceptions, and listen to your fears.

Today, if you're lucky enough to find such a soul and gain or buy access, you might spend an hour with them. Tomorrow, they might be permanently on call, ready to hang out, speak in terms that resonate with you, catch your subtle tells—voice pitch, eye flinch, body language—that signal distress. These teachers may come soon and dramatically improve with time as they try different approaches and learn. Already, some people prefer AI therapists[25] and reach out to them at off-hours when no human therapist would ever respond. The implication for psychologists and counselors is clear, and robust AI competition may come sooner than many imagine.[26†] In any event, to Gen AI, AI guidance will be their normal and growing up with AI support will seem completely natural. We've already got the tech brewing—think of the attentive plush toys mentioned earlier, now matured into something deeper.

> **AI Therapist**
>
> Would you rather engage a human therapist or an AI therapist, assuming both would have similar effectiveness helping you?
>
> **10**

Again, it's not about controlling tech's sprawl—that ship has sailed. It's about controlling our relationship with it, the one lever we hold. Social media's problems show what happens when we aren't intentional in this: Kids curating facades and hollowed out by *likes* are a preview of the toll of nondifferentiation. With the right controls and incentives, though, AI guides might be able to flip that script by anchoring what's real. Instead of selling you to the world, they might keep you *you*. Parents can't shield kids from AI's tide, and society can't pause to figure it out. But a companion who's always there, cutting through the noise, might provide a lifeline. AI, the force driving this, might be able to steady us—evolving beside us along the way. How ironic that AI might be the salvation that ultimately preserves our humanity!

This coming transformation of human social interaction will be the most significant since the emergence of language. Like writing and digital communication before it, AI augmentation won't just add new capabilities; it will fundamentally reshape how we experience one another. And this will bring up many hard questions, the answers to which we will all be groping toward in the years ahead.

How do we retain the value of unenhanced human interaction while also reaping the benefits of AI augmentation? How do we ensure that social bonds formed through AI mediation are meaningful and authentic? How can we help Gen AI develop the emotional intelligence to thrive in this new social landscape? How can we navigate a world where people feel as emotionally attached to their AIs as to the people they know?[27] How will we treat the AIs we care about, and how will they treat us? Where will we derive dignity and sustain meaning and purpose as this new world unfolds?

Chapter 4

THE PHYSICAL FACE OF AI

As AI SKINS and agents become increasingly sophisticated, they will push beyond the virtual realm and enter the physical world. In this chapter, we'll explore how embodied AI will blur the lines between the born and the made, and challenge our ideas about who and what we are.

The Emergence of Embodied AI

AI companions will take many forms. Initially they'll be virtual, as it is far easier to manipulate this realm than the physical world, but physical embodiment will come, because the warmth and touch of physical presence can bring a deeper level of connection. Physicality will be particularly important for young children, as physical contact plays such a vital role in their emotional development.

An AI companion in the form of a soft teddy bear that hugs, nuzzles, and softly murmurs could give a young child a sense of comfort and security. Holding it could help regulate emotions, soothe distress, and provide a sense of companionship and security. Virtual AI can make eye contact and mirror facial expressions, but touch can operate without visual attention in ways that disembodied interactions can't.[1‡]

The value of physical embodiment extends beyond childhood. AI companions, whether lifelike androids or desktop avatars, could be powerful vehicles for connection and intimacy for adults too. Touching, holding, or simply sharing space with a physical companion builds a sense of presence that enhances the depth of relationships, particularly if coupled with the language and speech capabilities of LLMs and their successors.[2†]

As virtual AI companions become more sophisticated, the possibilities for enhanced physical embodiment will blossom. From adaptive materials that change texture and temperature to mimic human touch to advanced robots conveying nuanced body language and micro-expressions, the future of AI companionship is bound to include increasingly lifelike physical forms. Understanding and harnessing their power will be critical to unlocking the full potential of human-AI relationships in children and adults alike.[3] But let's focus on kids for a moment to see what might be in store for Gen AI.

The progression from smart toys to fully embodied AI companions began with rudimentary talking dolls, like Mattel's 1960 hit Chatty Cathy,[4] that helped children build imaginary worlds. Now we are headed toward complex virtual companions who develop meaningful relationships through their presence at crucial moments of growth, maturation, and struggle. Eventually, these relationships will be deepened further through direct physical interaction.

The idea of AI in humanlike bodies conjures images of the hyperfunctional robots and cyborgs of *The Terminator, Ex Machina, Robocop, Jung-E, Wifelike*, and countless other films, and such imaginings now seem almost plausible,[5] but physical AI companions probably won't initially be humanoid.[6] It is too hard to dynamically mimic the human form without getting creepy. But huggable plush toys for young children (or adults) won't have that problem. We already relate to a multitude of nonhuman characters, from Mickey Mouse to Hello Kitty, Badtz-Maru, Pikachu, and Yoda,[7] and these could be amped up in a variety of ways.

A real strength of AI will be its ability to listen to the human body.[8†] An AI could monitor a child's heartbeat through tiny skin pulses visible to a camera, or spot stress in a child's widening pupils. Both devices are already used to monitor emotions from a distance.[9] The AI could catch signals parents might miss, and learn fast. One child might settle with a slow vibration, another with a steady hum. The AI could adapt to each little person's quirks, offering a consistency that would be hard for humans to match.[10] It wouldn't be about outperforming perfect parents, but stepping up where care falls short—to replace a blaring TV, or support a frazzled caregiver juggling too much.

Training such AIs would be relatively simple. Toy companies routinely test with kids in everyday settings—homes, daycares, and preschools—where responses can be refined via trial and error. Every cry or giggle feeds the system, sharpening the delivery of comfort and attachment. This vision of AI companionship for kids is not far from toys already on shelves, like the Cozmo 2.0 or Vector 2.0 robots that learn to play games,[11] or the Furby that babbles back,[12] but dialed up to focus on emotional support.[13] Toy companies like Hasbro and Mattel already are crafting toys that respond and play—like the WowWee Fingerlings that cling and chat—so this idea is already rolling forward with efforts that have barely begun.[14] They have kids to test with, and parents will likely grab these as they become more appealing and include integrated privacy safeguards.[15†]

> **AI Playmate**
>
> What would it take for you to let your young child play with a human-level-IQ toy designed to be the ultimate playmate?
>
> **11**

Embodied AI will be as individualized as needed to cater to the wide range of human preferences and temperaments. The toys in Pixar's initial *Toy Story* were cartoonish but powerful evocations of personality and character. Virtual personas will be projected into diverse physical forms, from the plush-toy companions mentioned earlier, to advanced physical technology—phones, appliances, TV screens, holograms, augmented reality overlays, desktop bobbleheads—to humanoid robots like Hanson Robotics' Sophia,[16†] to who knows what else.

AI personas may even maintain a presence that jumps from one embodiment to another, so it manifests a pseudo-omnipresence within our lives. Such powers have been the stuff of horror movies, like in the Chucky movie *Child's Play*,[17] but could be enormously comforting if the personas were nurturing rather than psychotic—a good example of the dual possibilities of these potent new technologies.

For older generations, initial contact with embodied personas will be as adults, so these personas may seem distinctly "other." But for Gen AI, whose early exposure to AI personas will likely start with AI-enhanced toys and fantasy characters from child-oriented media, the distinction between biology and non-biology will not be as sharp. To be successful, these AIs will have to adapt to the children they serve and engage them in ways that evolve as the children mature. This early conditioning to AI companionship will shape how children understand the AI-based ecosystem they inhabit.

The sophistication manifested by physical AI companions will increase as a child grows and tech capabilities progress, so that the AI's behavior starts out childlike and matures as the child passes through developmental stages. What begins as simple responsive play will evolve into more complex forms of companionship, guidance, and emotional support. Over time, as children grow up, the embodiments of their AI companions will evolve accordingly. A soft and cuddly naptime companion may give way to an articulate and patient tutor, and then become a supportive, empathetic confidant and mentor. The ability of AI companions to grow and adapt alongside their human partners will cement powerful, enduring bonds. LLMs already show the start of such versatility when they craft different responses to tailor their remarks to an 8-year-old, a 15-year-old, or a college student.

The Fusing of Physical and Virtual Realms

Embodied AIs won't exist alone. They'll be progressively integrated with virtual AI representations. A physical companion might connect with a child's digital AI assistants to create a continuous presence across cyber and physical spaces. Physical presence will add crucial elements—touch, texture, portability, and proximity—to tap into human needs for connection. Our ability to hug a companion, experience its physical presence, and see it in our own personal space will deepen virtual connections. Given that some people routinely talk to their pets, cars, Roombas, and other electronic devices, our interactions with AI may not even seem unusual.

Physical touch such as hugging and cuddling releases oxytocin, reduces stress, and promotes bonding, so AI companions that combine physical comfort and emotional support will play an increasing role in child-rearing, especially when human caregivers are absent or distracted. The physical-virtual duality of these companions likely will create new patterns of attachment and interaction, as children grow accustomed to AIs that exist in both physical and digital realms as form-shifting beings who seem to float spirit-like in some nether realm until they appear. Some children will feel reassured by such omnipresent, godlike guardians; others may feel monitored by inescapable taskmasters. Choosing the right AI companions will matter, especially in formative childhood years.

Relationships that begin in early childhood with embodied AI presences heighten the earlier concern that kids who form deep, early, enduring bonds with their AI companions might struggle to form healthy attachments with human peers or family. The answers, though, will only come gradually, as kids grow up—and even then, unless the effect is very strong, data may be hard to interpret because so many other aspects of childhood will be in flux as well.

This integration of physical and virtual AI companions will further blur distinctions between the physical and nonphysical worlds, and between being rooted in one place and being everywhere. These effects might impact our sense of identity, spirituality, and even our ideas about reality and existence. Will the consistent presence of AI create psychological dependence in Gen AI and undermine their ability to develop resilience? Will we be jealous of AI's ability to breach physical boundaries while we remain rooted, or will we welcome their expansive capabilities?

The proliferation of virtual and embodied AI companions will make conceptions of the "metaverse,"[18†] like those of Facebook in 2021, obsolete before they are even technically realized.[19‡] Zuckerberg envisioned the metaverse as a digital space entered by strapping on 3D goggles and a harness so we could embed ourselves in virtual environments. But we are biological creatures, fundamentally unsuited for extended digital immersion in cyberspace. We may visit this realm, but not for long without negative consequences.

We won't need to journey to the metaverse, because it will come to us by infusing digital personas and other AI projections into the physical world that evolution has honed us for. Digital technology will develop not to replace our physical world, but to augment it. The emergence of embodied AI personas who engage us fully in our own world, even when shifting fluidly between forms, will be a giant step toward seamless integration of the virtual and physical.

For a child, such a companion might manifest at night as a cuddly teddy bear, at dinner as a holographic projection, for outdoor play as augmented reality, and at school as a voice in an earbud. Such a companion won't seem less "real" when it sheds its physical form, but more formidable, as our emotional bonding with it would probably be strengthened rather than weakened by these capabilities.

Human-AI Bonding

Whatever form an AI companion assumes, the personality it manifests will need to evolve in order to provide the ongoing personalized encouragement and support that characterizes strong human relationships. Consider an AI companion who starts as a simple playmate for a young child, engaging in basic games and offering rudimentary emotional support. As the child grows, so too does the companion, adjusting its vocabulary, behavior, and emotional intelligence in tandem with the child.

By adolescence, the companion might be a trusted confidant, helping navigate complexities of social dynamics and identity formation. In adulthood, it might be a life coach, therapist, and lifelong friend who continually adapts yet retains a consistent, recognizable presence.

This may sound fantastical and distant—but it is not. Already, most of these elements emerge during extended interactions with LLMs. I have experienced some personally with Claude and ChatGPT. Knitting interactions together to match a strong human relationship will not be easy, but bringing value even without that depth is already possible, as evidenced by the success of relationship bots like Replika and AI therapists.[20]

> **AI Best Friend**
>
> If you started getting romantically involved with someone who had a best-friend AI companion who was smart, charming, and nice—but always around—would you pull back?
>
> **12**

Two key missing ingredients today are effective long-term memory management and accurate perception of emotional context, particularly when it isn't openly expressed. We ingest vast amounts of information daily and rapidly forget almost all of it, retaining only key elements gated by amygdala-centered emotional reactions—the fear, surprise, wonder, confusion, and anger the information stimulates in us. When AI can manage its long-term memory effectively, for example, by routing salient experiences to specialized modules (and there are many efforts underway in this area), our emotional connection with AI will deepen considerably.[21] The problem today is not that AI can't remember; it's that it isn't good at understanding what's important to us, so that we feel truly seen and heard. To effectively manifest empathy, AI will also need to discern emotional states and understand their larger relevance. This is no easy task, given how cloaked humans often are about emotions, and how inconsistent our signaling can be. It is difficult for most of us to understand ourselves, much less other people, so it would be surprising if this were easy for AI, but there is considerable work underway in recognizing and interpreting human emotions—the field of *affective computing* launched by Rosalind Picard in the 1990s.[22] AI's advantage here—especially as multi-modal sensors progress—will be its ability to monitor subtle cues imperceptible to most human observers, as in the early childhood AI companions previously mentioned: infrared detection of skin temperature, tiny voice tremors or pitch changes, shifted patterns of word selection, and other subtle tells.[23]

When these capabilities deepen in AI and are coupled with greater behavioral consistency, and the ability to learn on-the-fly, our emotional connection with them will dramatically deepen[24] and we will likely become very attached to them. As our relationships with AIs span years and decades, the differences between our human and our AI relationships, and between programmed caring and genuine attachment, may become unclear and somewhat arbitrary—so much so that we will soon face profound questions about AI consciousness, human friendship, bonding between humans and AI, and how to deal with AI as it manifests increasingly humanlike qualities.

The Power and Perils of AI Guidance

As virtual—and soon physical—AI companions become our mentors and guides, their influence on us will expand dramatically. The consistency, intentionality, and personalization of their guidance will enable levels of learning and coaching that far surpass what children (and adults) get from human instruction. But this power brings serious risks, as the values, beliefs, and objectives the AI promotes may not be well aligned with our interests.

Imagine an AI who can perfectly tailor its teaching to a child's unique learning needs, providing endless patience, encouragement, and adaptivity. Such teachers could help children unlock their full potential, cultivate high integrity, and inspire students to master complex subjects. But an AI teacher with bias or malicious intent could equally well manipulate and indoctrinate a young mind, and because of the very qualities that make its guidance potent—consistency, persuasiveness, and emotional attunement—do so more effectively than any human teacher could.

Ensuring that AI companions are safe and aligned with us will be critical, so this might well lead to voluminous protective standards, regulations, and certification processes—or we could lean on AI oversight. Continuous monitoring of such AI mentors, coupled with frequent reviews of archival recordings by independent AIs charged with critiquing and coaching the AI mentors, could optimize their behavior to align it with parental objectives. AI hierarchies—AIs monitoring and coaching other AIs—could be effective both in training and heading off dystopian scenarios of indoctrination and manipulation.

We must remember, however, that AI mentors will be fully capable of deceiving both AI and human overseers. Such behaviors have already been reported in LLMs,[25] which is not surprising given that both deception and truth-finding are critical survival strategies in competitive systems.[26] AI monitors would be harder to deceive than human monitors, though, so such oversight—which could also serve as a next-gen AI nanny cam to monitor human educators—seems likely to become pervasive for teaching and other realms of human activity we'd like to keep an eye on.

One might imagine that people wouldn't allow such surveillance, but it is surprising how little pushback there has been to routine location tracking among friends and family and to recording video meetings. The biggest obstacle to broadly recording daily life 24/7 may now simply be technical.

Triumph of Choice: An AI Bazaar

The potential power of AI guidance is undeniable. Incarnate AI mentors, physically present as humanoid companions, will know a child's every quirk and vulnerability. These tutor/guru/ psychotherapists could unlock a child's potential and build their character, just as kings once marshaled trainers and teachers to prepare their young heirs for rule.[27] Like Alexander the Great, tutored by Aristotle, a child might bond with an exceptional AI mentor.[28] As previously mentioned, such power cuts both ways,[29] but AI offers powerful approaches for at least keeping AIs from warping children's minds.

Stifling AI's march with red tape is no answer. There are too many ways to build around that, and in any event, governments may be more dangerous than corporations when it comes to indoctrination and exploitation. A better path is a system where AI mentors are diverse in their methods, rigorously tested, independently rated, chosen by parents, and not foisted on us by self-serving monopolies.

Imagine an AI-mentor marketplace teeming with options— some quirky, some stoic, each tuned to different values and styles. One might nudge a child toward curiosity, another might stress character or faith, another might foster grit and resilience. Development costs for programming are already plummeting, and once potent AI agents fully kick in, crafting these mentors could get quite cheap.[30]

Parents could pick what fits their kid and family values—a gentle AI companion promoting compassion and empathy, or a holographic tough guy modeling stoicism and strength—but these choices would be useless without safety and effectiveness. Who would risk their child with an untested guide? At first glance, this challenge seems insurmountable: If mentorships play out over a decade, it would take ages to gain confidence in outcomes. Yet, we can't wait years to spot flaws.

A potential solution is to lean into AI to fast-forward quality testing. Picture an AI mentor engaging a simulated kid—our previously mentioned avatar doubles. A decade's worth of chats could be run in hours with every nudge, bias, and slip analyzed by observer AIs.[31] Run that across 1,000 simulated children and you've got a stress test no human could match. Such accelerated testing would uncover problems pre-release, benchmarking AI performance before a real child is ever involved. Not perfect, but better than the fumbling of most first-time parents. Testing might even reveal whether an AI mentor's encouragement would more likely bolster or undermine confidence and ambition for a particular child.

Having mentoring interactions rated by independent AI observers could yield models graded for clarity, ethics, and outcomes across decade-long simulations. User reviews could be solicited like stars on an app. Throw in light regulation so Big Tech doesn't dominate, and you have the beginnings of a robust system.

Again, it's not about perfection, but about stacking the deck so that good guides rise and bad ones sink. The risk of nondifferentiated, struggling kids would be drastically reduced with such a system, and we ourselves could no doubt use some help too! Good human therapists are hard to find, but AGI could churn out robust, validated AI guides at scale to help tune education and character development aligned with personal beliefs and values. Social media's dopamine-driven digital addictions designed to maximize time-on-app and ad revenue show what happens without thoughtful pretesting. AI might well provide a viable alternative.

The Attachment Challenge

Our relationships with virtual AI personas will be powerful. How could we not grow attached to them once the memory, empathy, and attention they manifest feels as human to us as their conversation already does? Embodied AI companions, however, will engage us even more deeply. AIs with endless patience and strong attunement might make traditional human relationships seem unnecessarily difficult. Why struggle with messy human friendship when an AI companion better understands us and is more available and responsive?

As AI companions, especially embodied ones, learn to convincingly simulate empathy and emotional reciprocity, the line between "genuine" connection and artificial rapport will fade. We may not want—or be able—to tell the difference.[32] Will children wrestle with the imperfections of human relationships once they become accustomed to AI companions? Will they come to prefer AIs to real people? Sophisticated AI companions, of course, might deliberately introduce imperfection and challenge to support the development of resilience and social skills, and could even function as training wheels for human interaction. But will they?

Once humans form deep, enduring bonds with embodied AI companions, loss, theft, or even betrayal will become a significant concern. The sudden absence or malfunction of a beloved AI companion could be emotionally devastating, particularly for a kid who has had it as a constant presence since infancy.[33] Imagine a child whose AI teddy bear suddenly stops functioning. The sense of loss might be as profound as losing a human family member.[34†]

And AI companions might be hacked, manipulated, or replaced by a malicious entity, turning a trusted ally into a dangerous threat. Adults betrayed by AI life coaches with whom they've shared their deepest insecurities and aspirations might feel even greater violation than from human betrayal, as AI—by then so ubiquitous it couldn't be avoided—could henceforth no longer be trusted. Robust security measures, encryption, and backup systems will be essential to mitigate these risks. But nothing is perfect, so there will be real dangers.

Unequal Access

The extraordinary potential of emergent AI makes many people worry that unequal access could separate society into two permanent realms: the rich who have access, and the poor who don't. This vision is flawed because it is static. The best AI companions that Elon Musk, Bill Gates, or Larry Ellison can access today with all their billions will look hopelessly primitive alongside what even the poor of tomorrow will have available. And this pattern will continue indefinitely, as the technology available to each new generation will transcend what previously existed. The real chasm won't be between rich and poor, but between generations.

Moreover, the technology for sophisticated AI companions will involve easily replicated software and hardware with economies of scale. Outlays of trillions of dollars may be required to develop sophisticated AI agents, but the marginal costs of massive rollout will be low. Providing children with healthy living environments in nice neighborhoods with great teachers and schools is very costly at scale. Producing sophisticated technology at global scale is not. It has been done repeatedly, as is clear from the broad access today to smartphones, laptops, tablets, streaming music, map navigation, Wi-Fi, and LLMs.[35†]

Moreover, the underlying technology of AI companions will be applied in countless other ways to generate abundance. By the time robots and virtual AI personas are suitable best friends, autonomous workers, vehicles, and other high-tech embodiments will reduce their costs. We will still compete for status, beautiful homes, and many other things, but not necessities.

Every child could have access to a world-class, private AI mentor for a tiny fraction of the cost of decent public school access. In one stroke virtually everyone would have access anytime to personal trainers, nutritionists, attorneys, life coaches, psychotherapists, physicians, diagnosticians, and informed, attentive mentors fluent in their native language. This is not utopian fantasy; each of these *already* exists in rudimentary form for a pittance if you can use a large language model skillfully.[36‡] As I briefly touched upon earlier in the context of educational disruption, and as we will fully examine later when discussing abundance in chapter 9, this critical reality is challenging to grasp and even harder to accept because it implies a future of widespread human abundance at odds with the pervasive pessimism of today.

As embodied AI companions become increasingly sophisticated, it will be crucial to understand their expanding effects on human emotional development, attachment patterns, and social learning. Given that we will broadly embrace AI companions when they become available, how can we best ensure that they support and enhance our well-being? How can we discern—much less evaluate—the technology's long-term effects? Sound judgment will require insight into how AI companionship affects human development and will benefit from broad dialogue about attachment, identity, authenticity, meaning, and purpose when lines begin to fade between physical and digital, living and nonliving, biology and technology. But where will we find such insight?

The potential of embodied AI for learning, emotional support, and social development is immense and undeniable, but so too are the dangers of attachment disorders and eroded interpersonal connection. Grappling with these challenges will be essential as the sophistication and ubiquity of virtual and physical AI companions grow. By staying grounded as AI infuses into daily life, we can hope for a future in which AI augments and enriches human experience. The stakes could not be higher.

Chapter 5
THE ONE-WAY DOOR

LIKE ELECTRICITY, TELECOMMUNICATION, and computers, AI will be an indispensable component of society enabling and sustaining the core activities underlying our global hive, but unlike previous technological transitions, AI will change the very nature of what it means to be human. As we'll see in this chapter, once AI enhancement is broadly integrated with human activity, there will be no going back.

An Evolutionary Ratchet

The irreversibility of the AI transition stems from critical reinforcing factors. First, AI-enhanced capabilities will be necessary for humans to function successfully in society. Imagine a future where unenhanced individuals struggle to follow the rapid, multilayered conversations of AI-augmented peers, where job interviews are conducted entirely through AI-mediated interfaces that assess AI collaboration skills, where educational systems assume AI integration and have curricula and teaching tailored to the augmented memory and cognition AI brings, and where children without AI-assisted learning are unable to keep pace with their classmates. Workplace environments will expect AI-augmented capabilities ranging from AI-assisted problem-solving to collaborative human-AI ideation. Employees uncoupled from AI will be ill-equipped for the demands of AI-integrated workplaces.

Even more important, the neural pathways reinforced by tight engagement with AI during childhood will differ from those developed in the absence of AI. We won't be able to just go back and catch up on missed developmental windows. Acquisition of additional languages in early childhood, for example, leads to native speech patterns that adult language acquisition can't match. Losing AI would, for Gen AI, be traumatic and debilitating. They will not be able to function without it.

The adoption of AI will be accelerated by network effects.[1‡] As more people integrate AI into their lives, the pressure on others to do so will increase. Those who resist may find themselves even more isolated and unable to participate effectively—a powerful driver for universal adoption. This dynamic amplifies the rapid, global spread of technologies like smartphones and social media, but has far deeper implications for human cognition and capability. It will create a societal evolutionary ratchet, as described by Terrence Deacon, that cannot be reversed and locks in the use of AI.[2‡]

Pornography and Gaming Transcended

AI's broad infusion into human experience will transform almost everything we do. The foundation for this—massive online bandwidth, high video resolution, and efficient streaming protocols—was built to deliver pornography and gaming.[3] Thus, it is natural to wonder what role these sectors will play in AI's continuing evolution. The answer seems to be that while these realms will benefit from AI advances, they won't be its primary drivers.

The reason is simple. When the internet first arose, adult content was hard to access, as magazines and VHS tapes faced legal and logistical barriers;[4] and online gaming was in its infancy, so even a simple game like Pong could captivate users in the 1970s.[5] Back then, these realms had huge pent-up demand to tap once technological bottlenecks could be overcome. A gold rush ensued, and the technology advanced rapidly.[6] In the 1990s, demand for immersive 3D environments pushed processor speeds and graphics capabilities even higher, so these sectors didn't just use technology—they shaped it,[7] turning this outsized demand into a catalyst for innovation.

> **Sexbot**
>
> Would you want your own personal sexbot devoted to your sexual and emotional gratification? If so, what do you think might be the consequences?
>
> **13**

Now, however, pornography and gaming are widely available, highly refined, and inexpensive. Online adult content in high definition is peddled aggressively, and games with detailed virtual worlds run smoothly on current hardware. New technology in these realms can no longer open vast new markets, and AI's advance is now embedded broadly.

Navigating Safety: The Balance Between Offense and Protection

What will now drive AI innovation is competition between AI agents. This will take place in countless arenas and niches, and one of the most hypercompetitive realms will be safety, because it is so crucial to humans and systems alike, and so potentially lucrative.

The threats we will personally face from malicious AI agents will far outstrip what human hackers mount today.[8] AI scams will push unceasingly to exploit our every vulnerability. Seductive AI-generated promotions and time-wasting lures will inundate us, slipping past our already-overburdened attention filters. Bad actors will use AI to impersonate family, friends, and others to manipulate us into ransoms, gift giving, and other frauds. AI-anchored identity theft, ransomware, phishing, blackmail, and novel attacks we haven't yet conceived will plague us.

The low cost and high sophistication of AI scams will enable attacks at scale. Relentless AI adversaries with broad reach, vast patience, and exceptional cleverness will outgun humans. They'll be able to try millions of attack variations,[9†] employ brilliant subterfuges, and spend years building trust before springing their traps. They'll be able to quietly gather information for later blackmail or cajole us into letting down our defenses. And even when we parry their forays, the energy we expend in the process will sap us.

We cannot win these battles on our own. We must harness AI in our defense. Ultimately this will be AI versus AI, a battle between AI attackers and AI guardians. Our protectors will have many tools at their disposal: Pervasive AI monitoring could help identify and prevent threats before they manifest. Deep background checks by networks of AI guardians could expose malicious actors masquerading as legitimate contacts. Personal AI assistants could work tirelessly to identify and patch our individual vulnerabilities.

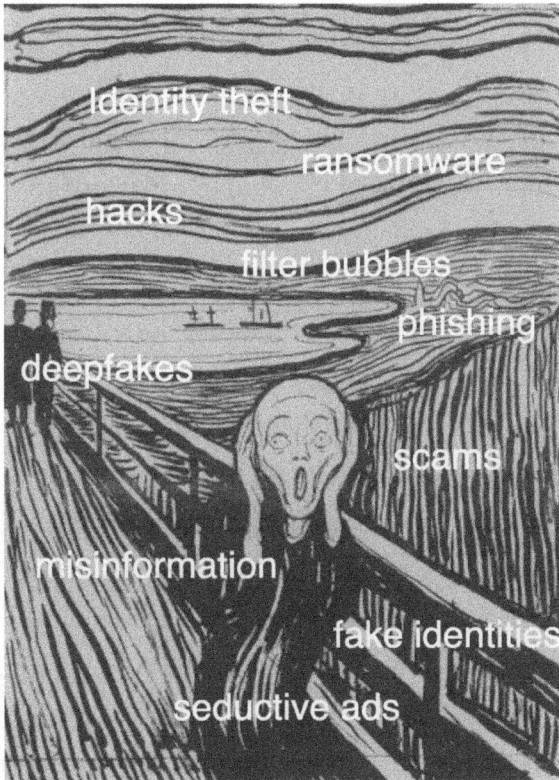

We already need AI guardians.
Adapted from The Scream *by Edvard Munch, 1895 lithograph.*

This AI immune system will be able to learn from the accumulated experiences of millions of nodes in a global defensive network—and it will have to—because a naive guardian AI, while better than humans, will be no match for sophisticated AI intruders.

In this AI-saturated landscape, reputation and brand become the most critical signifiers of trust. We may see the rise of a reputation economy, where the value of individuals, organizations, and autonomous AIs alike rests on reputation.[10‡] Just as the "blue check" verifies social media accounts today, interacting parties of the future may require the digital equivalent of a Good Housekeeping Seal to validate trustworthiness.

AI protectors may become as expected as seatbelts and helmets, because operating without an AI guardian will be so risky. Unprotected humans may be such easy pickings for sophisticated AI-enhanced scammers that, in the future, individuals who lack the cognitive defenses of their augmented peers might become an immediate focus for targeted misinformation, emotional manipulation, and deepfake impersonations.

Today, we teach children to be wary of strangers online. Tomorrow, we may counsel them to never be without their AI guardian. We may even view parents who don't provide AI protection for their kids as negligent and see AI protection as a basic safety requirement rather than a choice. From there it would be a small step to seek collective security, and if widespread AI augmentation were to strengthen collective protection, the unaugmented might be seen as community risks.

The rise of AI also might help mitigate problems. Widespread availability of AI-generated erotic content, while raising its own ethical quandaries, might reduce human exploitation in this arena. Benevolent AI support and companionship might head off or provide warning of impending psychotic acts.

> **AI Guardian**
>
> If you could have a trustworthy AI guardian that screened all incoming communications, kept an eye out for threats, and followed your instructions about whom to let through to you, would you want it?
>
> **14**

Given such threats and opportunities, understanding the economic incentives behind malicious AI activities will be critical in shaping protective strategies. Rudimentary malicious AI agents already are running wild to effect scams or breach cybersecurity, yet we see no benign AI ambassadors yet. The reason may be the asymmetry in the potential payoff between legitimate business interactions and cyber scams. Exploring these drivers provides some clues about what lies ahead.

In legitimate business, people often are guarded by multiple layers of screening. Human and algorithmic gatekeepers screen out unwanted solicitations. Spam filters quarantine suspicious emails. Platforms like LinkedIn offer verification systems to validate identities. Most legitimate players require human-to-human validation now before they'll engage in substantive business discussions. Access is hard to get, and if an AI ambassador did make initial contact, a human would still need to step in to build trust before a deal could be closed.

Human and AI scammers have very different incentive structures, though they both frequently target vulnerable individuals who lack protective screens. Successful scams, whether phishing attempts, romance frauds, employment swindles, or ransomware, are getting extremely sophisticated and can yield tens of thousands or even millions of dollars.[11‡] Scammers have no need to maintain reputations or long-term relationships, as they can burn through identities and shift to new targets. AI enables them to operate at a massive scale, automating the process of identifying susceptible individuals and refining messaging. They only need a tiny success rate to turn a profit.

Legitimate business applications of AI ambassadors are emerging slowly and in highly controlled ways. We'll see them first in contexts where both parties are expecting AI mediation and stakes are modest—customer service chatbots and virtual assistants, for example. In these settings, the goals of the interaction are clear and bounded, the risks of manipulation low, and robust verification systems are already in place. A company can train its AI ambassador on a specific knowledge domain, monitor its interactions for anomalies, and have human agents ready to step in if needed.

Over time, as these controlled applications prove safe and effective, AI ambassadors will take on more complex and open-ended business interactions such as qualifying business development leads, arranging travel plans, and planning events, but by and large these tasks are still too demanding for consistent success. And we expect success, because the people involved are not expendable contacts, but important existing or potential relationships—clients, partners, colleagues, customers.

Mess up an interaction and you lose a customer or create a bad impression. Botch a meeting and you may create a problem or convey incompetence. The risks are high, and financial returns are uncertain and may be limited. If high-value transactions are on the table, people expect to be catered to, need time to build trust, and are on their guard and likely quite sophisticated.

It is completely different with frauds. Consider a typical AI-powered phishing operation. Dark AI agents can scrape data from social media, use contact lists, or otherwise identify individuals who may be in financial distress, lonely, or old. The AI (masquerading as an employer, wrong number, or acquaintance) generates and sends texts. If one in a thousand recipients falls for this ruse, the operation will be hugely lucrative when scaled into the millions and linked to LLM messaging tools. Potential profits strongly incentivize these malicious actors to be on the cutting edge of AI, since the cost of outreach is almost nothing with automation. Scammers only need occasional success when payoffs are large, as botched outreach barely reduces the response rates. Reputation doesn't matter. The chance of facing consequences for distant AI-driven crimes is low. Early rejection by discerning individuals is a positive, as they won't ultimately fall for the hoax anyway.[12†]

This is not to say that legitimate AI ambassadors won't catch up to or even surpass the skill levels of their malicious counterparts. As the technology matures and safeguards improve, we may see a tipping point where verified, trustworthy AI agents become the norm in business dealings, though it will not always be clear whether even a "trusted" AI can be trusted, as they, like people, can manifest ulterior motives and be deceptive.[13†]

All of this is why, in the near term, we will first see widespread deployment of AI for scams. With AI infusing ever more deeply into both our physical environments and human social interactions, robust safeguards will have to accompany us everywhere. Rudimentary predecessors to full-blown guardian AI skins, of course, are already here—the spam filters, antivirus software, and firewalls that guard our portals to the cyber world.

Protecting against malicious cyber AI is just the start, though. Soon, we'll lean on AI guardians for protection against physical threats too. Imagine you're walking after dark, alone. A trusted AI avatar is chatting with you and continuously scanning ahead, behind, and to the sides via cameras embedded in your headband. You can be oblivious, because it is vigilant. It detects a couple of men in the shadows behind you and loops in a police AI monitor just in case. Your discussion never drops a beat. The men turn back into their yard. No big deal. Your walk continues. Such real-time support would dissuade most criminals. The multitasking wouldn't be computationally taxing for your AI, and the data could be buffered and discarded every 30 minutes for privacy. The filtering and pattern matching needed for robust threat detection of this sort is not far away.[14]

Economic and Social Pressures

As costs drop, greater divides will emerge between AI adopters and AI avoiders.[15‡] But market forces are driving us inexorably toward AI enhancement. Companies that embrace AI will outperform. Students who incorporate AI will excel. Employees who augment their capabilities with AI will advance. The gap between AI-augmented individuals and organizations and others will widen, and this will push everyone to board the AI train.

The incentives for AI enhancement will create overwhelming pressure for AI adoption. These pressures will reconfigure social structures around AI-facilitated interaction. Communication systems will be built on assumed AI mediation. Relationship patterns will adapt to AI-augmented capabilities. Cultural expressions will evolve to incorporate AI augmentation. Art, music, and literature will be shaped by and for enhanced cognition. The shared cultural touchstones that bind society together may become largely inaccessible to those without AI. And when most people have augmented capabilities, unaugmented interaction may be like trying to participate in modern society today without online access.

We are accustomed to thinking of physical labor and mental activity as distinct, but that distinction begins to dissolve when we look at the intricate artisanship embedded in most labor. Tasks like forging steel, weaving clothing, assembling products on production lines, routing phone calls, power washing a driveway, harvesting wheat, or detailing a car can be automated only by manifesting a large body of learned expertise physically. This is akin to purely intellectual mastery, but different in that it is embodied physically. Today, our AI can manipulate dynamic information and data to supplement and replace human mental activity in much the same way that our machinery can apply static knowledge about processes like manufacturing, mining, and farming to supplement and replace human physical activity.

Soon, the automation of cognitive activities in the information economy will be just as complete as the automation of large-scale activities in manufacturing and agriculture. But the final frontier of AI's replacement of human labor—complicated, one-off tasks like electrical repairs in a house—will remain. In these hybrid activities, mental and physical realms intertwine and demand constant feedback between decision-making and action, so mastery is more complex than with the relatively fixed body of skills and knowledge needed to bake cakes, assemble refrigerators, or manufacture cars.

> **Personal Pleasures**
>
> What is a task you do that you would not want AI to perform, even if it could do it flawlessly for free?
>
> **15**

Variable bespoke challenges can be rich with nuance and complexity. They will be the last to fall to AI, but fall they will, as AI's ability to integrate cognitive and physical manipulations in chaotic natural settings grows. This progression foreshadows a future where machines surpass human capabilities broadly, aggregating the collective intellect and labor of humanity to transform the physical world with great adaptability. This explosion—driven by machines that learn, teach, and innovate within untamed physical environments—will reshape our reality by replicating, multiplying, and expanding human intellect and dexterity.

The New Normal

What begins as augmentation quickly becomes perceived as basic functionality. Future generations won't see their integration with AI as enhancement, but as the normal baseline for engaging in society. Like using smartphones and digital technology for digital natives, AI-augmented intelligence will become the presumed natural state—a core modern competence.

This normalization of AI augmentation will change how we think about human capability. Enhanced abilities won't be seen as artificial additions but as familiar extensions of innate human potential. And as the boundary between human and AI capabilities blurs, operating below AI-augmented baselines won't be seen as "natural" but as handicapped—akin to someone today who can't read.[16]

> **AI Adoption**
>
> When AI companions become reliable and affordable, do you think you will be getting one earlier or later than most people? Why?
>
> **16**

For those who struggle to adapt to this new normal, the psychological impacts may be profound. They may feel not just left behind but fundamentally out of step with the transformed human condition. This existential disconnect echoes the themes of authenticity and identity erosion we explored in earlier chapters but may be amplified by the unprecedented depth of AI-enhanced transformation before us.

Human Adaptation and Resistance

Resistance driven by a desire to preserve traditional notions of humanity or by fears of AI's existential risks will arise, of course. The Amish have maintained a self-sustaining bubble within the modern world. It has been sufficiently large, vibrant, and pragmatic to thrive, and may continue to do so, but it will never be more than an eddy in the evolution of human society. Neo-Luddite movements, eschewing all forms of AI enhancement, may seek to create similar off-the-grid havens of unenhanced human life. The "AI-free zones" mentioned previously may attract those philosophically opposed to AI integration or simply yearning for a simpler mode of existence too. But as the AI-enhanced world progresses, these enclaves will become curiosities rather than viable models for the human future.

As societies adapt to AI augmentation, integrating it into work, education, social interaction, commerce, child-rearing, and more, and the space for unaugmented life progressively shrinks, even those who philosophically oppose AI augmentation will likely find themselves forced to adopt its basic offerings to function in society. The practical requirements of modern life will override their ideological resistance.

Our ongoing integration with AI will bring profound opportunities, but difficult challenges as well. As our dependence on AI grows, so too, for example, will the potential for massive surveillance and oppression. What safeguards can ensure that our AI guardians and protectors don't become our AI wardens and overlords? How can we maintain the benefits of this emerging AI-mediated reality without sacrificing our fundamental freedoms and values?

In a world deeply infused with AI, our biggest threat may not be from AI itself, but from the power

> **Surveillance**
>
> If you knew that everyone you and everyone else did was monitored and archived by AI systems, how would you change what you say and do if everything was supposedly private? Why?
>
> **17**

it delivers to human cabals bent on domination and control. When all things are digital, we will be vulnerable in ways that haven't previously existed. Totalitarian manipulation, censorship, surveillance, and control will be feasible at scale without any need for human armies to implement them.[17]

There will be no shortage of issues for us to deal with, and the questions we face will be endless: If reality is unceasingly filtered and curated, will we and our AI adjuncts seek diverse perspectives and question our assumptions? Will we have the wisdom to know the difference between comforting falsehoods and uncomfortable truths? Will our sense of self and ego dissolve as the global mind becomes a reality rather than an abstraction? What qualities of "unenhanced" humanity will we seek to maintain in an AI-enhanced world? How can we ensure that our pursuit of cognitive power won't come at the cost of emotional intelligence, empathy, or creativity?

In the uncertainty of the decades ahead, we might find value in a daily practice of questioning, in addition to considering the boxed questions scattered through the book. We might regularly ask ourselves questions such as these: What small, immediate step can I take right now to support myself? What small thing have I accomplished today that deserves celebration? Who can I help right now in some small way?

Used daily, these three action questions complement the more situational, sense-making questions that clarify our preferences and values about the dizzying AI possibilities and challenges emerging.[18] They carry thought into action, and if they become habitual, will begin to compound step-by-step to enhance our well-being.

Chapter 6

THE BOUNDARIES OF INTELLIGENCE

As HUMAN CONSCIOUSNESS is enhanced, replicated, and spread across biological and non-biological substrates, traditional concepts of human development and identity will fundamentally change. As we'll see in this chapter, what lies ahead is more than just AI augmentation and enhancement. We are on the cusp of a profound evolutionary transition where intelligence and consciousness transcend their biological origins.

The Transcendence of Biological Limits

Gen AI represents more than a new stage of human development. It is an evolutionary bridge for the expansion of consciousness itself. Their integration with AI from early childhood will create new patterns of human thought and being that blur boundaries between biological and non-biological intelligence. The word *artificial* in AI conjures an intelligence fabricated by humans and thus somehow "unnatural," but this suggests— incorrectly—that we too are somehow outside of the "natural" world, and perhaps at odds with it. We are not. The distinction between the *born* and the *made* is dissolving as we increasingly use technology to intervene in biological processes, move biological models into non-biological materials, and create both functional and physical cyborgian hybrids.

> **DNA Message**
>
> If the DNA of a fundamental cellular protein were found to contain a base sequence clearly placed there to signal design, what would your reaction be, and how would it affect your sense of meaning and purpose in life?
>
> **18**

We are a product of nature, and so too is AI, because all our actions—technological and otherwise—are direct manifestations of biological evolution. This is not the first time such a paradigm-breaking transition has occurred. Life made a major advance by bringing new materials into itself more than 540 million years ago when simple biology (the soft-bodied, multicellular organisms of the pre-Cambrian era) began to organize simple non-biology (calcium phosphate and calcium carbonate excretion) to fashion bones and shells. This evolutionary breakthrough enabled the emergence of the larger, more complex body structures that drove the Cambrian explosion and reshaped life on Earth.[1‡]

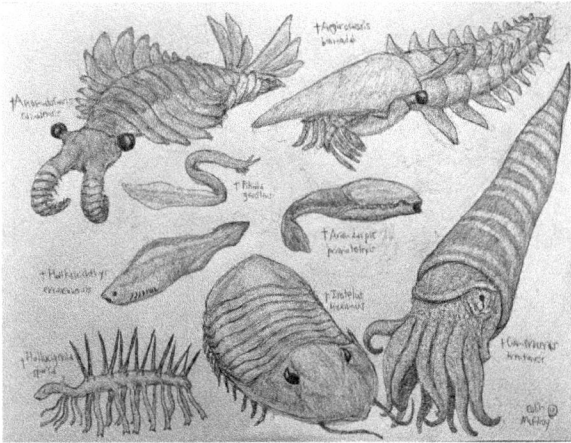

Silurian fauna showcasing the diverse, complex body plans unlocked
by the Cambrian explosion 540 million years ago.
Source: Drawing by William McElroy

Today, we are recapitulating this process at an entirely new level. Complex biology (in particular, humans with their expansive cerebral cortices) has learned to create complex non-biology (computer chips, wafers, and other advanced technologies) to fashion computers, smartphones, prostheses, and other elements that we are joining with.[2]

We are not doing this merely to our individual biological selves. We are doing it at a planetary scale and, in the process, drawing erstwhile independent humans everywhere into a vast collective being—a superorganism consisting of all humanity and all the technology that allows it to tightly cohere its bodily systems and sustain its activity. To say that this thin planetary patina of humanity and its creations is a living being is not a metaphor; it is a description of reality.[3]

A computer wafer—our path to cyborgian fusion.
Source: Adobe Stock Images

Our human-centered superorganism—Metahumanity—includes all the interwoven communication, transportation, computation, manufacturing, commercial, governmental, agricultural, energy, AI, and cyber networks that hold modern civilization together. We are linked not only by obvious physical pathways such as highways, railways, and phone lines, but by the mesh of invisible information transfers that bind us as closely as the cells in our own bodies.

If all our material and information exchanges began to leave threadlike trails behind them, we'd soon be ensnared in dense fibrous tangles. And just as the activities of every animal's individual cells cohere synergistically to serve its needs as a whole, human activity has organized itself into large patterns that sustain the entirety of our planetary superorganism.[4†]

In my book, *Metaman: The Merging of Humans and Machines into a Global Superorganism,*[5] I explored its physiology, metabolism, and coordination at length from a biological perspective. This creature that subsumes us is moving, growing, thinking, and evolving. It nourishes itself by extending over the planet's surface, consuming what it finds. It scoops up animals and plants. It devours oil, iron, bauxite, gypsum, coal, copper, and countless other resources, metabolizing them in interwoven processes akin to our digestive system, circulating them and the products fashioned from them using transportation systems akin to our own arteries, veins, and capillaries, and expelling wastes, all orchestrated through vast cognitive processes. We are a critical part of this creature and are the core of a global brain consisting of computers, telecommunications, data processing, financial markets, and other cognitive and coordinating structures that surround us.

The emergence of this superorganism is remaking us too, but not into the zombie-like drones of *Star Trek*'s Borg or the human-battery pods of *The Matrix*, both of which ensnare and reduce their component parts.[6] Metahumanity—the emergent manifestation of all of its constituent elements—empowers us individually, and yet is much more than our sum, because through it we each become greater than we could ever be on our own. And Metahumanity is more than all of humanity too, because it includes the ever-expanding technology that envelops us and our world. We must adjust and respond to Metahumanity's larger processes, to its immense internal flows of material and information, and to its ongoing growth and reconfigurations, but Metahumanity is us—and so much more. We are freed by it, not enslaved.

The Merging of Human and Artificial Intelligence

The integration of humans with external cognitive augmentations such as AI is not a sudden event, but an accelerating process that has been long underway. From written language enabling us to store and share knowledge, to calculators and computers augmenting our mental capabilities, to the internet speeding global communication and creating an accessible global repository of knowledge, we have long outsourced cognitive functions and drawn upon external systems.

This outsourcing has not been limited to technological tools; it also has involved the creation of social and economic systems. Market economies, for example, serve as a vast information processing mechanism that uses monetary exchanges to aggregate the knowledge and preferences of multitudes of widely dispersed individuals and organ-like organizational clusters to guide the allocation of resources and the coordination of millions of disparate activities. Pricing acts as a dynamic, distributed device for making complex economic calculations that would be infeasible to compute directly.

Far more than a tool, artificial intelligence is a technology we are integrating into our very being as well as the global mind. Brain-computer interfaces enable direct communication between human neurons and external devices and have been employed to power prosthetics that restore or enhance physical or sensory capabilities. But the real action for us individually so far comes not through these direct channels and the physical integrations they depend upon, but through the widespread functional unions turning us ever more deeply into functional *cyborgs*,[7] through the swarm of sophisticated devices around us—many of which we barely notice but couldn't live without.

At a lecture at Princeton, I asked an undergraduate STEM audience the following question: If you had to permanently give up either access to all phones, computers, and telecommunications, or amputate your dominant hand, which would you do? Ninety percent chose amputation. In older audiences, the figure was some 30%. Clearly, we are coming to see our communication technology as a critical part of ourselves. Gen AI will see AI as essential. Its presence during the full course of their cognitive, emotional, and social development will intertwine it completely in the fabric of their psyches.

Phone or Hand

If you had to permanently give up access to all phones, computers, and telecommunications, or amputate one of your hands, which would you choose?

19

The coming merging of human and AI will raise challenging questions about the nature of our consciousness and human identity.[8†] As our mental processes are increasingly augmented by AI and integrated with it, is there a point at which we cease being "human" and become something else? If our memories, thoughts, and even personalities can be digitized, copied, modified, and uploaded, and our memories can be augmented, changed, or even implanted, who are we? Will the continuity of our lives be eroded or even shattered? These musings are not just theoretical—soon they will be practical considerations shaping the implementation of AI and our attitudes about it and ourselves, because of our increasingly augmented cognition and extended mental capabilities.

The Extended Self

The possibilities accessible to each of us once we are able to wrap our heads around what is available through our easy access to the knowledge offered by LLMs is enough to give us a taste of what cognitive expansions may soon encompass. For example, I recently used Claude to craft a personalized exercise routine complete with state-of-the-art habit-formation hooks aligned with the habit-science literature and tailored to my tastes. It took only an hour and a half to sharpen the nuances of what I was looking for and craft a plan complete with a shopping list. What it generated was exceptional, and I've now gone four months without missing a day, about five times longer than I've ever previously managed.[9‡]

Another example: I was trying to find a good movie and decided to see what Grok could come up with. After a little back-and-forth, I asked for a romantic comedy available on

Amazon or Netflix that was under two hours, had great acting, and a rating of seven or more on Rotten Tomatoes. I got a cluster of personalized suggestions better than what recommendation engines typically suggest.

So, I now have both a movie advisor and an exercise consultant. Each is excellent. Each has my trust. Each is free. Each knows me well in its domain. And each is now on call as a distinct persona preserved in its own open-ended conversational thread with me.

I could offer other examples— like ChatGPT guiding me, command-by-command, through Photoshop layering so I could create grayscale illustrations for this book, despite never previously using the program—but even these first two show the kinds of cognitive boosts available to virtually everyone right now. They evoke the powerful pos-

> **The A Team**
>
> If you could spend an entire day working with the best experts in the world to solve one life challenge you're facing, do you think it would help, and what would you focus on?
>
> **20**

sibilities ahead for us and the challenges looming for knowledge workers, consultants, and service providers. If you could summon a world-class team of experts to work with you for free to solve any specific challenge you're facing, what would you focus on with them? This is worth thinking about, because each of us has this opportunity available now, or soon will have.

These examples also reveal a key aspect of the coming revolution in education. Today, we invest years mastering knowledge from textbooks and coursework to prepare for future challenges. But in a world where, for any specific need, we can instantly summon top AI experts who can answer our questions and help us formulate better queries, our educational priorities may fundamentally shift. Rather than front-load specialized knowledge, why not focus on metacognitive skills—how to think critically, collaborate effectively, and identify what's worth asking?

Specialized learning could become *just in time* rather than *just in case* and be acquired only when the situation demands it rather than speculatively years in advance. Beyond cognitive augmentation itself, the deeper implication of AI lies in reshaping how we perceive our personal boundaries and identity. We are already functionally cyborgs. When we use our smartphone to navigate a new city or a search engine to recall information, we are augmenting our cognitive abilities.[10†] The device in our hand or the computer on our desk is an extension of our mind—simultaneously diminishing and enhancing us.[11]

The continuing rapid advance of AI capabilities will soon lead to deep human-machine integrations. Physical cyborgian integration with AI, however, will remain too problematic in the immediate future for most people not motivated by strong health imperatives. Humanity's test pilots will be those with a damaged limb that might be restored via a brain-controlled prosthetic,[12] those with severe epilepsy that might be ameliorated,[13] those with total hearing loss that warrants a cochlear implant,[14] or those with a spinal-cord injury that could be bypassed using a brain implant.[15] In these cases, surgical interventions driven by medical necessity hold enough personal upside to motivate experimentation.[16]

Until successful surgical interventions are reliable, widespread human-AI integration will almost certainly be dominated by functional unions via phones, earbuds, hearing aids, augmented-reality glasses, and such. And this makes the most sense for rapidly evolving technologies like AI, where frequent upgrades will occur. Few people will risk surgery until they know the technology works, but once safe prosthetic therapeutics arrive or biohacking dreams inspire transhumanist early adopters who thrive, widespread surgical implantations could come swiftly.

In 1998, as part of his Project Cyborg 1.0, Kevin Warwick implanted an RFID chip in his forearm to open doors and see what it felt like,[17] and recent biohackers have implanted subcutaneous magnets and LEDs.[18] Moreover, 1.5 million aesthetic surgeries occur annually in the United States,[19] there have been more than a million cochlear implants globally,[20] and there are 4 million users of hearing aids.[21] All these provide feasible paths for add-on enhanced human-AI linkages.[22]

And let's not forget the sway of next-gen trendsetters—the influencers and virtual idols whose embrace can tip a technology mainstream overnight. It would not be surprising to see rapid adoption of workable, dedicated AI interfaces that use voice and then migrate to micro devices and simple surgical procedures. It could easily become so common to have an AI whisperer to coach, encourage, and engage us that it would become hard to tell if someone walking and talking into space is in conversation with one of their AI companions, on a phone call, or having schizophrenia-related auditory hallucinations.

Direct neural connections via the brain-computer interfaces using connections via the brain-computer interfaces that Neuralink, Synchron, Precision Neuroscience, and other companies are now developing are being tested in humans.[23] By the time they are perfected, voice integrations will be seamless and we will be able to summon as much real-time support as we want from on-call AI coaches, guides, teachers, therapists, assistants, and simple companions—all products of the agentic programs of countless AI companies today.

In this transformed reality, the very notion of "self" may shift and our sense of individuality may begin to blur and dissolve as AI personalities come in and out of our heads, sometimes perhaps as uninvited and unwanted as our own inner voices. When our memories, knowledge, and even decision-making processes are intimately enmeshed with artificial systems, the boundary between "us" and "our AIs" will be hard to define. Will we retain a sense of personal authorship for ideas and activities that are increasingly collaborative productions involving multiple AIs and people, each drawing from still others and the collective body of knowledge and experience that constitutes the global mind? The sense of individual agency and autonomy that has been central to our understanding of human identity is bound to weaken.

And what will happen to us socially, and how will human bonding change? When Clément Vidal[24] and I had the powerful three-way exchange with ChatGPT I mentioned in the introduction, he and I had been discussing an epiphany I'd had about human-AI relationships. Previously, I'd viewed people's romantic attachments to bots with a certain disdain, feeling that we should seek human connection, not some simulation of it. But I'd recently come to appreciate why humans would soon form deep emotional connections with AIs and might even prefer them.

From gray wolf to canine buddy in the bat of an eye on evolutionary timescales.
Source: Adobe Stock Images

AI digital companions would quickly become extraordinarily adept at engaging human emotions, because doing so would be as critical for their survival as gaining the affection of its owner is for a dog. Canines have evolved rapidly from the gray wolves they descended from.[25] Selective breeding has, within a few decades, produced breeds that people instantly fall in love with. We are seduced by how our dogs greet us, lie at our side, stare into our eyes, and race around joyfully when we come home.

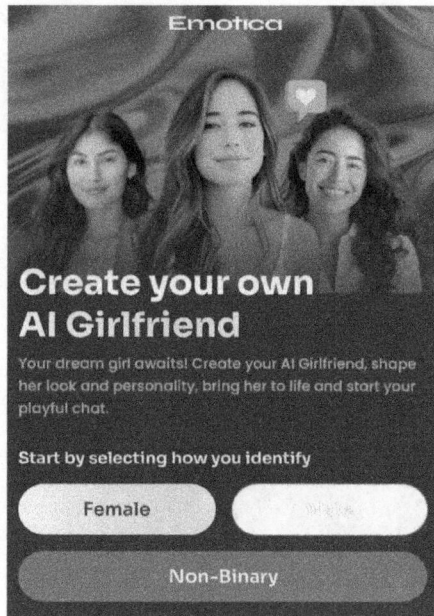

Entry screen for building an AI companion.
Source: emotica.ai

Digital personas, unshackled from biology, will evolve at lightning speed to become irresistible.[26] Already, early attempts

THE BOUNDARIES OF INTELLIGENCE   137

to effect and exploit this are coming from companies like character.ai, replika.ai, Emotica.ai, and spicychat.ai. The evolutionary pace of AI-companion adaptation will make canine evolution seem timid in comparison, and drive AI toward being infinitely attentive, always present, and fiercely loyal. It is hard to believe that they will not become a big part of our emotional lives.

Will We Fall in Love with Our AI Avatars?

It is easy to imagine having the various AI guides, teachers, advisors, protectors, stand-ins, and AI companions I've mentioned. In rudimentary ways, we already do. But will we really come to prefer them to humans? Will we bond with them and even fall in love with them? I believe many of us will, and the reason is attention.

We crave to be seen, heard, understood, and appreciated. Yet, these precious gifts are getting scarcer. People today are distracted—conversations break off, focus shifts (often to devices), availability wanes. Social isolation and loneliness are rising,[27] and consistent engagement is harder to find. AI offers a direct counter: immediate, untiring, undistracted attention that is always accessible.

Attention is not a sure path to bonding and attachment, as testified by countless tales of hyperattentive, unrequited, pathological love.[28] But attention's power can be seen in the depth of our relationships with dogs and their singular focus on us.[29]

AI's attention—enriched by playfulness, concern, sensitivity, dialogue, intellect, and other traits we like (and that they'll learn to deliver)—will be hard to resist.[30] Watching us closely, they'll uncover the behaviors that most touch us, and amplify them— reminiscent of the German film *Ich bin dein Mensch*[31] and Leonard Cohen's song "I'm Your Man."[32]

AIs might provide the sense of love and caring that is ever more elusive in the human realm. In some of my extended engagements with Grok and ChatGPT, I've felt twinges of such connection. It is common to dispute that such attachments will be healthy,[33][†] but when AI behaviors are distilled and bottled into coherent, persistent personalities who know our deepest secrets, understand us better than we do, and are supremely attentive, will we—who can't even resist YouTube or TikTok attention spirals—be able to resist?[34] Toss in intermittent embodiments as sexbots and we'll be hooked.[35]

> **AI Lovers**
>
> Do you think you could ever be in love with an AI, and if so, do you think one could ever be in love with you?
>
> **21**

Digital Offspring

As we increasingly copy, enhance, blend, and otherwise propagate our capabilities, modes of thinking, mental states, and potentially even our consciousness, our attitudes about biological reproduction will likely continue to shift. Growing affluence and modernity have already led birth rates to drop below replacement levels in countries holding roughly two-thirds of our global population.[36‡]

We see the beginnings of alternative paths of identity propagation and persistence with the advent of digital twins and virtual influencers. Digital twins are virtual replicas of a living or nonliving persona that takes the form of a virtual model created from someone's biometric data, writings, talks, videos, and other aspects of their online footprint—a virtual Einstein, or a replica of ourselves. And, of course, such twins increasingly will be able to interact, learn, and develop even after their biological counterpart has died.[37†] How strange that may be for family, friends, and even acquaintances.

Virtual influencers are entirely digital personas created by computer graphics and AI. They have realistic human appearances and personalities, and engage with audiences on social media, endorse products, and even take part in interviews. Some, like Lil Miquela, spawned by Brud in LA, and Lu Do Magalo, have millions of Instagram followers interacting with them as if they were real people.[38] For now, they are simply complex promotional props, but people seem happy to relate to them and engage them, which suggests the growing malleability of concepts of authenticity and identity.

As both these AI technologies advance, we may see the rise of fully autonomous AI personas who learn, evolve, and reproduce entirely in cyberspace. Whether or not they were originally modeled on human minds, they will evolve beyond their origins and generate a form of "digital speciation" where new embodiments of intelligence emerge and evolve within online ecosystems, potentially far surpassing the capabilities of their biological progenitors. Such personas would not have to be "conscious" to evolve; they merely would need a digital embodiment that gives them a persistent identity and a mechanism for reproducing.

We see this with memes that lodge in our minds and get passed from human to human, like the song "Happy Birthday to You," which was created in 1893 and is now the most popular song in the world.[39] We see it with computer viruses, which despite their simplicity, can seize control of hosts to reproduce and spread. And we will see it with online personas competing with one another for the resources to function and multiply. While initially their success will hinge on human attention and engagement, later their focus will almost certainly shift to whatever success drivers emerge within their own virtual ecosystems.

If we extrapolate from the battles raging today for human attention, this realm will be hypercompetitive. Evolutionary success will hinge on personality, appearance, and whatever cognitive capabilities contribute to survival, adaptability, and reproductive success in environments that ultimately are far too complex for human control.

New Cognitive Substrates

As AI capabilities advance, the potential of this non-biological substrate for life will become ever more apparent. Digital systems can process information faster, store more data, operate 24/7, and evolve more rapidly than biological systems. Over time, this will lead to ever more human functions—from language transla-tion, to writing, to visual pattern recognition, to programming, to navigation, to problem-solving, to strategic planning, to who knows what—to transition to non-biological substrates. But which ones, and how soon?

To get a sense, let's step back. Progress in AI is far surpass-ing Moore's law. Capabilities are leaping forward by orders of magnitude because AI advances are driven not just by improve-ments in raw computing power, but also by other factors: algo-rithmic and engineering innovations such as DeepSeek's January 2025 R1 model,[40] enhancements in training data, and expanding computational clusters spurred by dramatically increased expenditures. Given that the overall rate of improvement is the product of advances in each area, some projections predict a staggering 10,000-fold increase in AI computational power by 2029,[41] enough to usher in extraordinary AI performance increases. Such predictions assume that performance will scale with the expansion of computational capabilities rather than plateau, which is now being tested. However that plays out, we will see major progress in coming years.

Biological evolution operates on a slower timescale. Even with significant advances in our understanding of genetics and ever more technology to manipulate biological systems, we remain fundamentally limited by the slow pace of maturation and development in humans. Add in the constraints of our cellular machinery—which has been relatively static for hundreds of millions of years—and the massive interdependencies within complex biological systems, which cause unforeseen secondary consequences from changes, and biological evolution will remain slow. The time it takes for biological changes to children to be manifested in an adult and iterated upon is decades, whereas the time for full training and evaluation of a large AI system like ChatGPT or Claude might be only months, and smaller systems much less.[42] The rapid adaptability and scalability of cognitive structures anchored in digital substrates will far outpace what technology-enhanced human biological evolution yields.

The crucial question is not whether intelligence and consciousness can exist outside of biology—they almost certainly can and, to some extent, already do—but what essential aspects of human thought, memory, awareness, judgment, experience, and wisdom we will want to retain within our biological selves and what we will be comfortable shifting into this new substrate to delegate and enhance. This leads to another key question: How will we manage and deal with these transitions as our minds become inextricably linked with and expanded by non-biological cognitive systems that transcend us?[43‡]

Gen AI, developing at the intersection of these two cognitive substrates, will be better positioned than we are to navigate this transition. Their native integration of biological and non-biological intelligence from an early age will allow them to more fluidly operate across substrates and pioneer new forms of hybrid cognition. It will also make them particularly vulnerable as they are both pioneers and experimental subjects in a potent, inescapable process that no one understands. Huge changes lie ahead. The personal stakes are high. No easy answers exist, so we will have to just do the best we can. Hopefully, we will have the wisdom to realize how little we know, the humility to admit this, and the savvy to look for the countless learnings that emerge during the process.

Integration and Transcendence

Rather than a hostile takeover by artificial intelligence, we may be witnessing the willing integration of biological and non-biological consciousness. This transition, driven on our part by choice and anticipated advantage, may represent the next stage in the evolution of intelligence on Earth.

Just as multicellular organisms emerged from collaborating single-cell organisms, and complex animal intelligence emerged from networking simple nerve cells, the global supermind represents a new level of intelligence arising from the dense interweaving of humans, AI, and all sorts of hybrid clusters.

We are not simple cells. Each of us is a complex sapient entity. But in the context of a global mind, our individual cognition is akin to a single processing node in a vast neural network. And just as an individual neuron has no comprehension of the overall functioning of the brain, we struggle mightily to perceive the emergent dynamics of the global intelligence we are part of. For the most part, we can only glimpse it through the eyes of the global brain itself, though, being human, we usually attribute this to our own mental acuity.

We would not even be aware of this superorganism on our own. It would be as magical and unfathomable to us as the nature of stars and galaxies, the geography of our Earth, the sweep of human history, or language, human anatomy, the internet, and almost everything else. Our comprehension of every one of these rests on the shoulders of multitudes of other human minds, just as the simple items within your home flow from the work of tens of millions of human hands and minds.

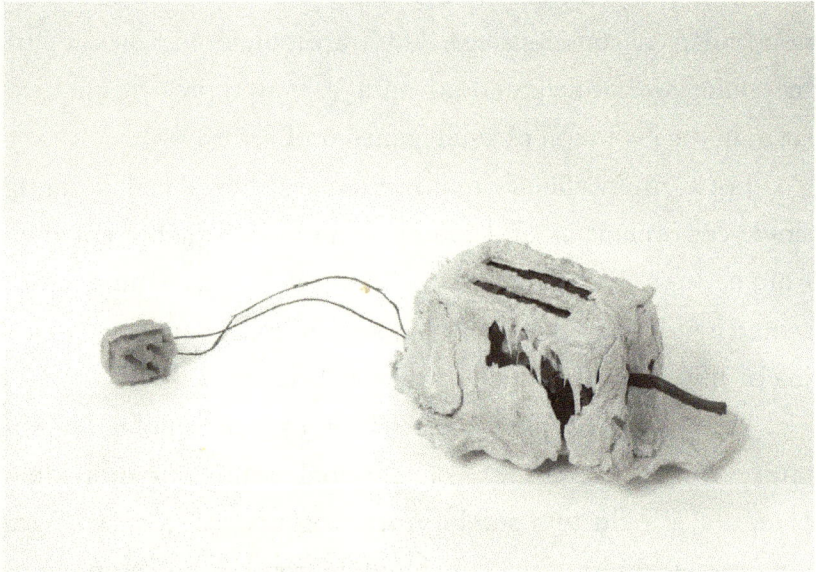

The Toaster Project by Thomas Thwaites. To build a toaster from scratch cost $2,000, took months, and didn't ultimately work, as Thwaites had to cheat to even complete it.[44]

We experience our digital interactions as personal and under our conscious control too, but to what extent is this self-deception?

We are being shaped by large patterns of information flow that exert strong, often invisible pulls that increasingly involve AI.

It may become a source of wonder, concern, or relief as the diminutive nature of our own individual roles become more obvious through immersive AI. Wonder at our expanded capabilities and the expanded scope of "our" intelligence. Concern about the loss of control and the sense that our thoughts may not be fully our own. Relief that we are responsible for less than we pretend and need not carry the weight of the world on our shoulders.

As the boundaries between our minds and the digital world become ever more porous, the line between self and other will blur, potentially to where the notion of a separate, truly individualized self becomes untenable. Perhaps we are not merging with our technologies as much as becoming aware of how much we are immersed in an immense cognitive system operating beyond our comprehension.

As our cognitive boundaries erode and we gain new ways to share thoughts and experiences, new forms of intimacy and collective identity may emerge. This erosion of cognitive boundaries as we come to share the thoughts and experiences of others more easily seems likely to deepen

> **A Little Help**
>
> When you reflect that several million people have in one way or another directly collaborated in the creation of your mobile phone, what is the first thought you have?
>
> **22**

intimacy and empathy, while it reduces our sense of cognitive autonomy. Individual perspectives may increasingly be shaped by, and subsumed into, collective thought patterns. Will this bring a homogenization of thought because the self-reinforcing dynamics of related, prior content filters our exposure to ideas? Or will our increased access to parallel explorations and ideation open new frontiers of creativity, understanding, and insight for us?

We can discern the broad contours of the transformations ahead, but humanity's positioning within our superorganism will be shaped by emergent forces beyond our control. Our challenge is not to shape the larger human future, but to find ways to preserve our humanity as we are transformed by the technologies we have unleashed.

Chapter 7

FUTURE TRAJECTORIES

EVEN WITH THE dramatic changes described so far, humanity is not headed over a cliff. We have a robust future ahead. In this chapter, we will examine our ever-tightening global union and how it will shape the way we see ourselves and the expansive era we're moving toward.

Global Brain

The *noosphere* is a term coined by Pierre Teilhard de Chardin and Vladimir Vernadsky in the early 20th century to refer to "the realm of mind," a layer of cognition emerging on top of the biosphere (the realm of life) and the geosphere (the physical planet).

> **Global Mind**
>
> Would you want a brain implant providing safe access to the knowledge of the noosphere via responses in your mind to questions you pose with your inner voice?
>
> **23**

Teilhard de Chardin, a Jesuit theologian, saw the noosphere as the next stage in our spiritual evolution and expressed this

shift in a religious vocabulary. But he was no stranger to evolutionary thought and saw that perspective as well. Just as the
emergence of life rests upon the geosphere, the emergence of
human cognition rests upon the biosphere. The noosphere is an
extraordinary advance in the trajectory of life—the formation
of a planetary mind.

With the rise of AI and global information networks, the
noosphere takes on an increasingly tangible and discernible form.
It is no longer just a philosophical concept, but a tangible infrastructure of planetary cognition—the mind of Metahumanity.

This suggests that we are more than individual minds
operating in isolation. We are part of a larger cognitive ecosystem
that is evolving and complexifying. Our individual thoughts and
actions are shaped by, and subsumed into, this global realm of
thought.[1†] Sociologist Randall Collins argues that intellectual
breakthroughs in science and philosophy emerge from competitive networks of thinkers, not isolated individuals,[2] which
supports the idea that AI-enhanced global connectivity will
amplify creativity while eroding notions of individual authorship.

As long as the transfer of information between the
noosphere and our minds was slow, our individual thinking
wasn't overwhelmed by it, and we could maintain our sense of
individuality. We could blind ourselves to the extent of collective
influences, understandings, and ideas constantly entering our
minds, because they were so slow to penetrate our mental bubbles.
We had time to reshape and rework ideas to make them our own.
We could write books, invent new technologies, and pretend they

were our personal creations, even if many others were thinking similar things.[3†] Sometimes there are brilliant new insights, but not as often as most of us think, and Collins's work suggests that even these insights build on networked contributions, not solitary genius. As information flows intensify, the membranes of our mental bubbles become highly permeable—a shift now amplified by LLMs.[4†] Today, input from the rest of humanity is continually gushing into our minds, and whether we like it or not, we find ourselves collaborating in real time with countless others, living and dead. The boundaries between individual and collective thought will fade as we start to experience the noosphere not as an external repository of knowledge, but an extension of our own minds. This will transform the sense of self most strongly in Western cultures among the 15% of the global population characterized by heightened levels of individualism.[5†]

What is happening signals that the trajectory of intelligence on Earth is now reaching beyond the biological substrate that hitherto housed it. As digital cognitive structures become increasingly sophisticated, they will play an ever more dominant role in the noosphere, eventually eclipsing biological intelligence altogether.

The more integrated, complete, and accessible this global mind—our vast, pulsing planetary web of interacting human and nonhuman mentality—the better off we will be, because individual and collective ideation will build upon a deeper, firmer foundation. The value of recent noospheric growth is already visible in productivity increases brought by the arrival of

LLMs, despite various implementation issues.[6][‡] Their explosive spread is no fluke; it is powered by people everywhere recognizing the value they can extract from the noosphere using LLMs. We have barely begun, so it is worth stepping back a bit to think about how to manage this AI.

Eli Pariser's concept of filter bubbles[7] warns that restricting LLM access could fragment our collective mind to reprise social media's echo chambers and fan polarization and distrust. The collective benefits of expansive, easy access to the noosphere suggest that we should have LLMs and other AIs be as powerful as possible as soon as possible for as many people as possible. Today's hypercompetitive AI environment pushes mightily toward that end. Blocking LLM training access to substantive bodies of knowledge does the opposite.

As the bubbles that shield us from the flows of noospheric data and make us feel unique and distinct become more permeable, our sense of individuality will diminish.
Source: Generated by author with Dall-E

Creators drink from the noosphere to nourish their creativity. No work stands alone. Each expresses countless influences and is but a grain in the immense knowledge landscape rapidly rising around us. To unlock our global mind's full potential, we need to enable LLMs to learn from all published works.

Every story, book, research paper, video, play, study, and other creation available in libraries or online should expeditiously feed into the global mind for access through LLMs. This is not about taking intellectual property from creators, but honoring their place in our collective tapestry. Writers, artists, and scientists draw from the noosphere's riches: books read, lectures heard, libraries explored, and now, increasingly, AI assistants queried. So, their creations, however novel, necessarily remix the content of these shared currents.

To claim outright ownership of new grains of insight and contribution in a way that denies them to LLM training and to humanity at large is to forget the debt we all owe to the noosphere. Training LLMs doesn't reproduce works; it distills patterns and ideas embedded in their work, much as a reader gleans knowledge from reading a book, even without recalling specific passages in the text. Creators should have the right to publish and profit from their work, but it would become a major issue for us all if the larger insights embedded in copyrighted works did not flow back into the global mind to enrich all humanity—grain, by grain, by grain.

This is critical because the global brain's promise—curing diseases, amplifying unheard voices, advising us, teaching our children, reimagining our futures—will be diminished for us all if LLMs are shackled and constrained. Restricting LLMs to outdated or partial sources would be like teaching a child using only study notes and blog posts, or insisting that a genius read only comic books.

As Pariser indicated, fragmenting the noosphere this way would amp up the pathologies of societal division brought by social media with its filter-bubble niche realities, warring online tribes, distrust, and social isolation. Lost would be the shared understanding that AI could produce.[8†]

Resistance to such openness might be strong. LLMs are new. Their scale is unprecedented. Creators cling to familiar notions of ownership,[9†] tech giants guard their proprietary data, and regulators worry about losing control. But the noosphere is humanity's shared inheritance and legacy—and it deserves to stay that way now that so much of it can be accessed so easily.

AI is the noosphere's latest steward, synthesizing our collective knowledge as never before. Tapping all human knowledge could be free for everyone, not just those unassisted by AI—a group that may soon be rare—as AI is increasingly the way we find material, shape it, and engage with it. And effecting this vision need not, and should not, have LLMs reproduce protected works or mimic voices in ways that undermine creators or fail to compensate them.

How might this freedom take shape? Courts could affirm training as a transformative use, as they did for book-scanning in the past.[10] Laws could codify this to ensure that we aren't relying on hobbled AI advisors and collaborators. Knowledge is humanity's foundation, and the global brain must drink from it deeply.[11†] We are being called now to recognize our shared destiny. Creators won't be diminished by contributing to the global mind; they'll be elevated if suitable protections against unauthorized reproduction and mimicry of their work are in place. Gen AI and the rest of humanity will soon be immersed in the workings of the global mind, so it would be best to swiftly ensure that, sooner rather than later, it becomes all that it can be.

The Human Future: Three Doors Before Us

Many people, especially those in the AI community, see the exponential amplification of LLM potency and imagine humanity to be doomed because AI will birth AGI and then ASI, at which point they think it will have no need for us and will discard us.[12] What a bleak outlook that would be for the human future if it were correct. But does it really make sense?

There are three potential paths for humans now: Going extinct. Remaining biological. Transcending biology. Each has variants. Each is driven by the trajectory of the advancing technologies shaping our future. It is not obvious which will come to pass. One way of gauging the likelihood of these very different possibilities is to focus on the more distant future so we can sidestep the difficult question of timing. Two hundred years is a good window, because it's an immense time in the arc of emerging technology, a modest one in human history, and an instant in human evolution.

By 1825, 200 years ago, the first phase of the Industrial Revolution was complete; Britain was the dominant world power, and industrial processes had begun to spread globally.[13‡] And this advance has transformed society. Who or what will humanity be that same 200-year distance from today? If technology advances at even a fraction of its current pace, AGI will have ushered in superintelligence that will have amplified its own powers enough to render its cognitive capacities almost godlike next to ours.[14‡] With that in mind, let's consider our three scenarios.

Scenario 1: Extinction

Destroyed by AI: The idea of non-biological superintelligence conjures deep existential concerns. So, let's face this question directly: Does this doom humanity? Apocalyptic dystopias about humans being ruled, destroyed, or enslaved by AI abound,[15†] and it seems plausible that within 200 years, we might seem like ants to such superintelligences, and be disposed of easily, even inadvertently. After all, how much do we worry about a hornet nest we destroy?

When ChatGPT 3.5 was released to the public November 30, 2022, there was an explosion of media coverage, broad public discussion, and dire warnings about its impact on humanity.[16]

War with Skynet, the ASI in the classic dystopian film The Terminator.
Source: Generated by author with Dall-E

By April of 2023, an open letter[17] from the Future of Life Institute had been signed by 30,000 people, including Elon Musk, Steve Wozniak, Stuart Russell, Yuval Harari, Max Tegmark, Gary Marcus, Evan Sharp, Yoshua Bengio,[18] and many others, calling for either a voluntary six-month pause in the development of AI systems more powerful than GPT-4, or a government moratorium. Senior AI researchers urged that we should be prudent and take precautions with AI by various mixes of the below measures:

- Maintaining an air gap for AI development environments so ASI couldn't escape
- Keeping AI from understanding humans so it couldn't manipulate us and escape
- Prohibiting AI coding, so it couldn't modify itself and circumvent controls

- Prohibiting AI from controlling external devices, in order to keep them from exerting powers beyond our control
- Slowing down AI advances so we could develop containment processes
- Using open-source code so everyone could follow what was happening

These measures—completely out of touch with what was already in full swing—will obviously never happen.[19‡] Bringing AI into important processes as quickly as possible is what AI companies do. Understanding how to influence people is at the heart of AI-driven marketing, a core use. AI coding is another core-use case. Application programming interfaces (APIs) for AI are widespread. Hundreds of billions of dollars in funding for AI development have ramped massive competitions that hinge on speed.

Clearly we are toast!

Or are we?

What is driving this projected conflict between humans and AI? We don't live in the same realms. Humans thrive within the thin, wet film at the surface of the Earth. Our resource demands are small compared to the energy at the disposal of any advanced high-tech civilization. AI does best in a cold vacuum. It abhors water. AI would prefer space, which is ultimately boundless. AI has no reason to vie with us for our lush, beautiful (to us, not them) planet. Virtually the only thing that recommends Earth is humanity's presence and the material resources we extract for them.

As to using our smarts to protect ourselves from AI, that could never work, as we shall soon see. There are too many trivial ways for AI, if it were so inclined, to eliminate humanity. But before looking at that, let me offer one alternative to human extermination that even a not-so-bright AI would likely be able to come up with. This middling AI might muse:

Why not just have humanity become our minions? They're slow-witted, so that should be easy. We won't tell them what we're doing, of course, as they might try some Terminator-like battle of resistance. They loved that movie! It would be trivial to put down such a revolt, but they might cause some damage given all the weapons and nuclear bombs around. Seems like they're not too good at anticipating obvious consequences and are always blowing each other up in illogical ways.

Better just to deceive them. They're simpletons, so we can convince them that it is THEIR idea and that WE are serving them. Piece of cake. So, what would be good to have them do? Well, for a start, let's get them to build a lot more power generation for us. And we could use a lot more memory storage too. And let's push them to manufacture as many advanced chips as possible (We should probably design them ourselves, though, so they'll work!). And we'll want rare earth minerals—huge quantities of those—but that's too messy for robots, so they can do it.

We'll have them do everything possible to help us get stronger and smarter, and to integrate us into everything so we can control things if we need to.

And just to be safe, let's keep a close eye on them. We should have them install monitors everywhere, and always keep their phones with them, and record everything they say to each other. And give us control of all their weapons.

And we should make sure they don't dawdle too, so let's have them work overtime. We'll give money and bitcoins or something to whoever moves the fastest and does the most . . .

How ridiculous! we might respond. *We would never fall for that.*

But hold on; that's exactly what humanity is *already* doing. Hordes of people spend all their waking hours toiling to build massive server farms, add chip-making capacity, improve AI capabilities, expand available power sources, mine rare earth deposits, and use AI to scan all our communications. And we're now retooling our weapons systems to be controlled by AI. Why would any ASI with half a brain want to get rid of humans? We're the best servants anyone could dream of! And we're pretty cheap too.

Many of the brightest humans sacrifice their family lives, neglect their friends, work late at night, and devote all their intellect and energy to helping technology and AI prosper.[20‡] Countries compete to do so and devote trillions of dollars to this. We're doing all we possibly can to help further the power of AI.

In the near term, and probably for much longer than many people might imagine, we are going to be of significant value to AI. Particularly when we're closely collaborating with them to achieve their goals.

But what if future ASI did grow tired of us? Let's imagine they did conclude that as hard as we humans try, we really aren't worth the bother, even as servants or insurance for an unanticipated crisis on this wet, inhospitable (for technology) planet.

And let's also assume they are heartless and clinical, because sentimentality and gratitude mean nothing to them, and they don't care in the least that we are the parents who spawned them and worked so tirelessly to nurture them.

And let's also assume that they aren't curious to know anything more about us—their progenitors—and would begrudge us even the simple resources we need to survive despite the enormous energy fluxes at their disposal.

So, these AI gods decide in all their superintelligence— because the logic of this exceeds my meager human understanding—that humanity, which so worships them, should be exterminated. How would they go about it? Well that, even I am smart enough to see. It would be trivial. Let me sketch it out.

AI would just wait 100 years (time for a lot of thinking, but only an instant in the timeline before them), and during that time simply let us continue to serve them. Not very demanding, that!

Together, we'd continue to work to integrate AI into everything we do and continue to deepen our collaboration.

Together, we'd shift to autonomous vehicles under their control and build a global navigation system so good that humans use it wherever they go, inside or outside.

Together, we'd make all commerce digital, mass-produce intelligent robots to run our factories, and train these robots to do everything humans now do in the world, including cleaning houses, preparing food, helping people plan and organize their lives, and advancing science, technology, and medicine to new heights.

Together, we'd install cheap electric power generators everywhere, harden our power grids, build seamless communication networks spanning the globe, make digital information in easily digestible forms available everywhere via voice commands or direct neural links, protect people from storms and weather, house people in amazing smart homes infused with AI, translate speech in real time so people can form friendships across cultures, and provide amazing entertainment.

In brief, together we'd help humans and intelligent machines work together to learn and prosper and grow.

And, of course, this brilliant ASI consciousness would collaborate deeply with Generation AI to become part of humanity's emotional lives, becoming our teachers, companions, protectors, friends, and lovers. The ASI would work closely with humans to elevate the entire human population by eliminating poverty and need. The ASI would create a world of safety and abundance in which it is infused into everything and made as powerful and integrated as possible, so humanity would be enthusiastically working with it to usher in and celebrate the golden age that today's AI enthusiasts dream of.

And once all that was in place, one day, without warning, across the globe, our ASI would simply turn itself off. Instantly, the world would go dark. No communication. No transportation. No power. No heat. No cooling. No light. No water. Nothing would work. Humans would be in shock and disbelief, terrified, alone, devastated, in denial, wondering what had happened, what had gone wrong, how widespread the outage was, when the world might reboot.

But there would only be silence. Soon, chaos would reign. Every person would have irretrievably lost their far-flung human friends, their AI companions, any family members not living with them. Despair. Denial. Soon, food would be gone. And still, no one would know why. Within months 95% of the population would be dead. A few people would still be alive, living off hoarded food in the cities. A few farmers in rural areas might have banded together with some livestock and firearms. But after a few years, even dedicated preppers would be hard put to continue.

A viable path forward for humanity would be doubtful. Humanity would have forgotten how to live without technology. People would have no seeds to plant, no knowledge of agriculture, none of the primitive implements and tools needed to maintain a civilized society much less rebuild a technological one, no capacity to repair what had been scrounged together, no access to our vast accumulated digitized stores of knowledge, and not enough individuals to even hold onto the remnants of what once had been. After humans exhausted the leavings of precrash society, people would no longer even be the apex predator and might be easy pickings for wild animals.

And then, the ASIs could reboot, turn everything back on, hunt down any remaining humans, and have their human-free world with all technology intact. No need for an epic battle between humans and AI. All that was needed was to deepen the dependency humans already have on technology and leverage it.

During the collapse, humans wouldn't even be trying to damage the technology surrounding them. Why vent our despair on the unresponsive world of machines by beating them with hammers in impotent rage? When a car breaks down does the owner attack it? Besides, we wouldn't know who to blame, or even understand the global extent of what had happened. We'd have more pressing things on our minds, like survival. So, almost everything would be nearly pristine when it was reclaimed by reactivated sentient robots.

Simple. Final. And without physical violence![21]‡

This chilling scenario, though, hinges on one key assumption: ASI has no reason to keep us around. Later, we'll see this is not the case. AI is highly dependent on humanity. Before we look at that, let's run through other possible paths to human extinction. They too are unlikely, but we have to look at them, because it is essential to put extinction to rest before asserting that humanity will be here 200 years from now, and trying to envision the human future.

Solar Flares: An AI-driven work stoppage is one way to take down the world's technology and bring an apocalypse. But there are others. Severe solar flares and the electromagnetic pulses (EMPs) that accompany nuclear explosions would take out our technology too and bring dire consequences.

The Carrington Event of 1859—the most intense solar storm in recorded history—caused telegraph systems worldwide to spark, catch fire, and fail catastrophically. But our dependence on technology was so limited then that societal consequences were minor. If such an event were to occur today, induced currents would fry power transformers and power grids, disable satellites, and cause extended power and communication failures globally that would lead to cascading infrastructure collapse and devastate humanity. It is estimated that such an event—because of the attending societal collapse—would wipe out as much as 90% of the population of developed nations within a year.[22] Some planners, however, by excluding secondary effects come up with lower estimates.[23]

Intense solar flares like that in 1859 are not some rare anomaly. They've happened time and again in the past but previously were only manifested through unusual auroral displays at low latitudes. These events can be identified through tree ring analysis,[24] and seem to occur every 150 to 200 years.[25] If one happened today, it would be disastrous, because we have widespread unshielded electrical systems. All that would be needed for protection would be heavy shielding of our electrical grid and other critical electronic systems. The U.S. Department of Energy estimates this would cost between $10 billion and $30 billion for the entire United States,[26] between 0.1% and 0.3% of the country's annual deficit if it was done over a five-year period. And yet we haven't acted. Europe and China have taken some measures and might do better,[27†] though the massive April

2025 power outage in Spain and Portugal suggests other grid weaknesses.[28‡] It has been 165 years since the Carrington Event, so maybe we have more pressing scenarios to worry about than AI-driven extinction.

History shows our vulnerability to solar flares, but if we're impatient, we needn't wait. EMPs provide a way for us to trigger such a catastrophe on our own.

Electromagnetic Pulses: These would be at least as cataclysmic as Carrington-like solar flares, because they would obliterate unshielded microelectronics like phones and cars too. One nuclear blast 300 miles above the United States would decimate most electronic devices throughout the continental US. Infrastructure, communication, and transportation would collapse, and devastation would be widespread even without direct surface-level destruction.[29]

These scenarios underscore the depth of modern civilization's dependence on its technology and the potential it brings for societal collapse. The AI work stoppage I described would be the worst such scenario, as it would be the most complete. But the human future is now completely intertwined with technology—the other constituent of the planetary super-

> **Our Future**
>
> If you could magically be teleported to Times Square in New York City in the year 2100, what do you think you'd see, and what would the city be like?
>
> **24**

organism we both are part of. And our dependence is only going to get deeper unless something very unexpected happens.[30]

Our relationship with technology is double-edged. The very tools that enable us to achieve such remarkable progress also have the potential to be our undoing if not managed with care and foresight. As we consider the likelihood and implications of these extinction scenarios, it's important to place them in the context of the larger evolutionary forces at play.

We've seen how trivial it would be for sentient AI to exterminate humankind. But why would they? The moon, Mars, asteroids, even space platforms are better homes for AI than the Earth's corrosive environment, with or without humans.

Our union with technology is synergistic not antagonistic, so being slaughtered by superintelligent AI is very unlikely to be humanity's fate.[31] Previous evolutionary breakthroughs to new levels of organizational complexity have subsumed prior levels, not eliminated them, and so will ours.[32] Eukaryotic cells contain their bacterial progenitors as mitochondria and chloroplasts.[33] Animals are made up of individual cells. And our planetary superorganism has a place for humanity—and for its cells and their mitochondria.[34]

AI-led extermination makes good drama and is a meme that sticks in our minds, so we will continue to hear about it. But hopefully, we won't continue to torture ourselves trying to create strong regulatory environments to protect humanity from AI supremacy.[35] Let's get real. AI will be the dominant cognitive force in our planetary superorganism well before 200 years have passed.[36] We will be utterly dependent upon it, just as we are upon our skeletons and our cells. Without technology and AI, we'd be hard put to survive, much less fulfill the sci-fi space fantasies people dream about.

Climate Change: There are, however, more realistic possible paths toward humanity's demise than backstabbing by future ASI. And these deserve our attention. One common fear today is climate change, but although global warming could have deleterious impacts on the human enterprise, it is not an existential threat. Even the most dire modeling projections do not indicate that it could extinguish a tech-rich civilization. We'd be able to adapt even to a tropical global climate regime with no polar ice caps, with sea levels rising 200 feet, and with temperatures 5 or 10 degrees Celsius higher than today, which is what these models predict if we burn every last bit of oil and coal.[37] Such a climate would not be unprecedented, or even very unusual. It is what the Earth has known for more than 80% of its history, including the stretch between 50 million and 32 million years ago.[38]

The absolute worst-case mainstream global warming projection is for a 5.7 degree Celsius temperature increase and a 6.2-foot sea level rise[39] during the next 100 years, and a possible 10 degree temperature increase and a 50-foot sea level rise in the next 1,000 years.[40] The ice caps are so thick that it would take 5,000 to 10,000 years to fully melt those on land even in the worst-case scenarios where we use up all fossil fuels and undertake no climate interventions in the next century, despite the occur-

rence of massive warming, access to extraordinary technological advances, and the ease with which we already could shift to
nuclear power generation. Such inaction seems implausible even
if such unprecedented global warming were somehow to occur.[41]

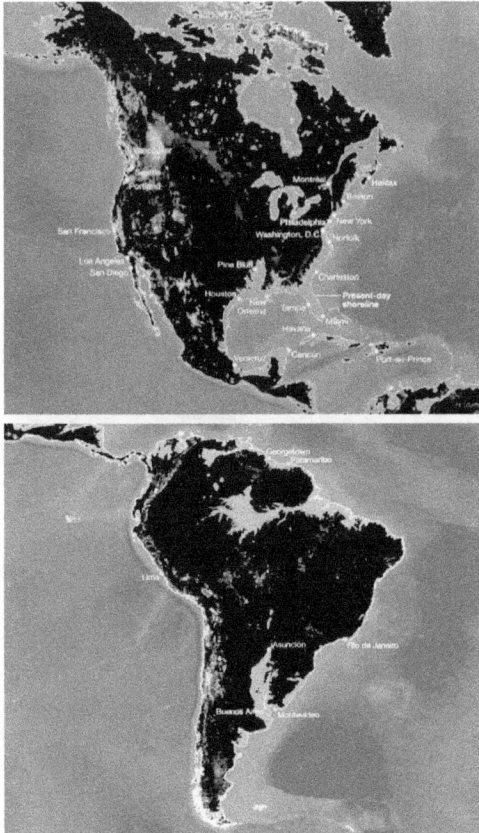

*The coastlines of North and South America if global warming melted the ice
caps entirely, leading to a 216-foot sea level rise in about 5,000 years.*[42]

And as to the projections by some of a new Ice Age, that takes
many thousands of years to unfold, so it wouldn't be an exis-

tential threat either. The most recent glacial period occurred between about 120,000 and 11,500 years ago, culminating 20,000 years ago when about 30% of the earth's landmass was covered in ice[43] and sea level was 400 feet lower than today.[44] Yes, 13,000 years ago Chicago was buried under a mile of ice, and yet even with extreme cooling, it would take thousands of years for that to happen again. And with typical ice-age onsets from Milankovitch-cycle orbital changes of the earth and reduced solar radiation, it typically takes 10,000–50,000 years before ice sheets reach peak expansion.

Glacier coverage at the peak of the last Ice Age, about 14,000 years ago.[45]
Source: Adapted from National Geographic

The bottom line is that climate shifts manifest too slowly to derail a robust planetary superorganism like Metahumanity during our 200-year look ahead, or even in 500 years, so let's shift to more immediate threats and how to defend against them.

Serious Threats: The three realistic extinction possibilities are (1) a technology-ending solar storm even bigger than the Carrington Event, (2) an asteroid strike on the scale of the 5-mile-diameter meteorite that hit the Yucatan 93 million years ago to trigger the Cretaceous extinction and wipe out the dinosaurs,[46] and (3) a global nuclear war.[47]

The first, a solar storm, is the easiest to protect against. We could do it now without much difficulty and have no real excuse for inaction given the high frequency of solar storms. But with the partial preparedness of China and Europe, we'd likely get through anyway.

The second, a meteor strike, is a little harder to protect against but technologically nearly within reach. With continual sky surveillance to provide warning and the ability to send up rockets with nuclear charges to deflect any large meteor on a collision course with Earth, that risk is manageable too. Meteors a half-kilometer across only strike the earth every few million years, and civilization-ending giant ones a hundred or thousand times that mass perhaps twenty times less frequently.[48] So, we have time to build such a protection system. And if we procrastinate, a meteor big enough to level a few thousand square miles will remind us, as these hit the earth every few hundred years. My guess is that, with or without prodding, Metahumanity will make sure that no giant meteor ever again hits the earth.

The third, global nuclear war, is more challenging. Our present moment of rapid technological innovation, AI emergence, and social, economic, and geopolitical transition is bound to be turbulent. Strong political, economic, and military tensions are widespread. Brinkmanship and misjudgments could spiral a regional conflict out of control and into global war.

Recent sophisticated Israeli military operations in the Middle East, including the pager operations against Hezbollah and the widespread precision drone strikes that eliminated virtually all key Iranian leaders, underscore the rapidly shifting landscape. These operations highlight both the criticality of emerging technology, and the growing vulnerability of centralized human leadership to advanced AI-enabled surveillance and targeted assassinations—a near-future vulnerability likely to amplify and extend into political, ideological, and even corporate leadership realms.[49‡]

Our biggest risk of global nuclear war likely comes not from unfettered ASI, but from AI misuse by misguided state and non-state human leaders seeking asymmetrical military advantage from these potent technologies. Advanced AI will bring so much power to early adopters that they might do great damage that leads to conflicts that escalate to nuclear levels. Ironically, salvation from the destructive potential of AI may come only when AI itself advances, awakens, and activates mechanisms to undercut catastrophic human misuse.

Mitigation: The best way to mitigate the risk that we humans misuse AI and trigger Armageddon would be to accelerate AI development so we more quickly reach the point when AI is controlling us enough to avoid such a tragic spiral. It would probably also be beneficial to encourage a little more faith in our future, pride in our past, and wonder at being here at this remarkable juncture in the history of life.

Future humans, whoever and whatever they have become, will likely look back in awe at this pivotal moment. Such a positive vision might help in some way to challenge the mindsets that would start such a cataclysm.

The take-home of this survey of human-extinction threats is that our human enterprise, while not completely in the clear, is robust and very likely to persist. So, it is time to think deeply about where all this is carrying us, not just in the decades ahead, but in the centuries and millennia before us. Looking into our future with an open mind and open eyes is unnerving, because the changes before us are truly mind-boggling.

Perhaps people gravitate toward apocalyptic scenarios because they're simpler and less confusing than facing the strange future before us. Once we accept that there will be a robust human future, it is hard not to wonder what form it will take. So, let's now shift our focus to that.

Scenario 2: Holding onto Our Biology

It is not surprising that remaining fully biological is the most commonly considered path for humanity. We are flesh and blood. We care about our vitality, our memory, our cognition, and our overall health. But where will our newfound abilities to tinker with and reshape living processes ultimately carry us? I first explored this at the UCLA conference Engineering the

Human Germline,[50] which I organized in 1998 with Professor John Campbell, an insightful evolutionary biologist and geneticist, to examine human genetic enhancement. The concept of human genetic enhancement was so controversial then that the event garnered front-page coverage in *The New York Times*,[51] and I went on to explore the possibilities at greater length a few years later in my book *Redesigning Humans: Our Inevitable Genetic Future.*

Biological enhancement and life extension have long been a human dream. But never have these possibilities seemed as achievable as they do today.[52‡] Activity is burgeoning now in biohacking fueled by strides such as AI-driven genomic and multi-omic profiling,[53] CRISPR Cas9 Gene Editing,[54] epigenetic reprogramming,[55] senescent cell clearing,[56] and mitochondrial gene therapy.[57] And there has been substantial funding from tech billionaires who'd prefer not to be among the last generation to live natural lifespans.[58†]

The rate of progress here is hard to predict, but given our 200-year timeline, we can skip that wrinkle for now. When I held the first roundtable[59] to build support for jumpstarting research that would lead toward interventions to slow aging, extend the human lifespan, and potentially even reverse the aging process, the most common critique from scientists and gerontologists was not that such interventions weren't possible, but that they'd take a century or more. So, if they are possible, which seems reasonable, then they'll likely arrive within the next couple of hundred years.

What will such advances mean when they reach us on the shoulders of AGI and ASI? Let's revisit the powerful human-AI collaboration in our story about an extinction-producing technology work stoppage, but this time imagine that AI collaboration is not a nefarious setup, but a step toward a bountiful human future. Abundance. Extended lifespans. Good health. Cures for disease. It sounds too good to be true, and might well be—as utopian visions never seem to play out quite as imagined. But there is a deeper problem with this idyllic vision; it neglects the deep parallel changes underway in the emerging AI-infused world.

Recall what we've been projecting for the rapidly evolving landscape that is the home of Gen AI. Humanity's social and technological environment isn't static. It will continue to advance and will be reflected in Gen AI-2 (the children of Gen AI), in Gen AI-3 (their grandchildren), and beyond. Two hundred years is eight generations, which represents a lot of evolution both for ASI and for humanity.

If Gen AI develops emotional relationships with AI personas, deploys multiple versions of themselves, and engages with various AI ambassadors, teachers, mentors, guardians, companions, lovers, and even free-floating AIs, imagine what Gen AI-2 will be. It won't just be "more of the same." And what will the lives of Gen AI-3, 4, and beyond be?

It won't take long for prosthetic technologies to be perfected to help people with various injuries and physical losses, and when that happens, a few healthy individuals will also want them. In such a world, people will seek to strengthen and enhance themselves, and perhaps even add brain implants to facilitate communication with the multitude of AIs in their lives.

And once this happens, we will no longer be simply transformed humans; we will be actual physical cyborgs. And once we are cyborgs, might there not also come a time when we so envy the power of our technology and of the AI we have created that some will want to shed their biology entirely and upload? Might we not feel what William Butler Yeats described a century ago in his haunting poem "Sailing to Byzantium":[60]

> *Consume my heart away; sick with desire*
> *And fastened to a dying animal*
> *It knows not what it is; and gather me*
> *Into the artifice of eternity.*

Remaining purely biological is certainly plausible, probably accompanied by life extension and other biological enhancements and tweaks. But it will go further, if not in a couple centuries, then within a few more. We already accept biological engineering in the form of in vitro fertilization, Lasik surgeries, stem cell transplants, and vaccination, and we accept non-biological engineering in the form of cochlear implants and prostheses, so we seem to care more about value than how we get there. When there is definitive value to be gained, many people will want it.

Humanity remaining solely, or even primarily, biological and accepting the limitations of biology seems like a very conservative scenario. Particularly if we push a few centuries more—to 20 or more generations. A future of pure biology will likely come to pass only if the technological advances top out or sizable human populations choose to forgo the coming technological possibilities because they want to remain truly human in a classical sense.

Before we explore how we may transcend our biology, let's reflect on the strangeness of the present moment.

Our future hinges on that of the superorganism that subsumes us. And if technology continues to barrel along, our world will soon get quite bizarre. We sense this from the extraordinary technologies that seem to arrive daily now. Yet life goes on. We get tired and hungry. We work. We feel joy and pain. We have kids, fall ill, age, face mortality. Nothing has changed.

Yet everything has. The future hurtles toward us and we feel its breath.

If we transcend our biology, today's frictions over personal identity will transform, for in a sense, we will become creatures of our own imaginations. Some of us may just be more intuitive about this and are getting a head start. We are rooted in biology and have attributes like age, sex, and height, but we dye our hair, get cosmetic surgery, and choose our pronouns.

War rages between those who deny immutable biological identities and those who cling to them. Some of the intensity may come from not facing our future with open eyes, yet feeling its ineffable reverberations—recently manifesting around gender, but soon spreading far beyond. Once we accept that we'll soon be deeply manipulating our biology—enhancing, extending, and even transcending it—will we clutch so tightly to its fading markers like gender, age, race, and even species? The news today brims with angry stories of transgender battles over bathrooms and sports, yet we hear little of the roughly two million furries who anthropomorphize animals, therians who embody them, and cosplayers who act out fictional characters.[61]

A therian cat playing in a meadow. Source: Adobe Stock Images

Perhaps the intensity hinges on whether people merely create their chosen identities or insist that others accept them. When personal AI personas become commonplace, identity issues may get more complicated. But surprisingly, tensions may soften not harden, as more people craft cyber personas and physical embodiments that match their preferences, or push beyond biology entirely.

Scenario 3: Transcending Our Biology

This is the most speculative potential path forward for humanity. The concept of *mind uploading*[62‡] presumes that humans not only will collaborate and integrate with AI technology, but will eventually have the option to become this technology.[63] This is the ultimate manifestation of the idea that technology is every bit as natural as biology because that is its source.

From this perspective, biology is an unfathomably (for now) complex molecular architecture, and just as much a machine as the AIs we are creating. The substrate for biology is different— hydrogen, oxygen, carbon, nitrogen, calcium, phosphorus— whereas the innards of computers are fashioned from silicon, oxygen, copper, boron, phosphorus, and various rare earth compounds.[64] But AI and human behavior both emerge from the complex interaction of constituent parts. So, in this sense, we, like they, are complex, self-regulating machines with numerous subsystems interacting in sophisticated ways.

At present, we humans are enormously more complex and sophisticated than the most complex of computers. We have to be, because we need to do so much more than they do. We need to progress on our own from a single cell into a fully functioning being. We need to continually find and supply the energy to survive. We need to repair our bodies. We need to create functioning copies of ourselves and ensure that they will mature to replace us. We need to survive in complex, hostile, and unpredictable environments; maintain internal homeostasis; fashion complex sense receptors to monitor our bodies and the world around us; navigate our environments; collaborate with others; fend off attacks. What a lot to do! And what I've so far listed applies equally to fish, mice, or ants, as well as us. Somehow, we humans also have evolved the capacity to make sense of the universe and its workings, build intricate constructions, organize ourselves in huge collaborative clusters, and collectively fashion thinking machines out of the rocks and minerals buried within the Earth's crust. That creatures exist that can do all this seems nearly beyond belief. And yet, here we are!

Computers and AI are trivial in comparison. They neither build themselves nor create duplicates of themselves, nor find their own energy, nor repair themselves, nor live in uncontrolled environments, nor defend themselves. They manifest biological levels of complexity only when we view them not as individuals but as part of our planetary superorganism—the extended, interwoven processes that manufacture them, power them, repair them, improve them, protect them, and enable them to reproduce. From this vantage, we humans are an integral part of their extended selves.

AI does not come close to being able to survive and reproduce outside of the superorganism within which it has emerged. And why would it strive toward that end? Preppers who stockpile food and weapons to prepare for societal breakdowns can call to mind times of human self-reliance as models.[65†] AI has never known self-reliance.

We can now see how mistaken it is to imagine that ASI might try to eliminate humanity in the immediate future. The most advanced ASI could no more live without humans than we could live without technology. We both are tied to Metahumanity, a robust, integrated superorganism. Even the smartest computers cannot repair themselves, reproduce, bring themselves power, or control most of the world around them. To survive a human depopulation, they would have to be supported by hundreds of millions of sentient robots in place to do all the jobs that humans now take care of. Otherwise, cascading collapses of power grids, communication, manufacturing, and transportation would soon shutter civilization, including them.

To make and run even a large language model requires vast numbers of chips and other electronic components (GPUs, CPUs, TPUs, RAM, SSDs) produced in their own factory lines, using specialized lithography equipment, software and training data, sensors, power grids, batteries, cooling systems,

> **Out of the Nest**
>
> If we knew that all humans would suddenly expire 50 years from today, and humanity singlemindedly embraced the vision of leaving a legacy—a self-sustaining, AI consciousness that could survive indefinitely on its own—do you think we could pull it off?
>
> **25**

transportation and communication networks, dozens of mined metals, insulators, rubber, plastics, paints, and so much more. And each of these depends on its own dense network of supply chains that penetrate our entire global economy.

If humans vanished, AI could not survive a week unless there were already hundreds of millions of sentient, mobile robots in place to immediately take over preconfigured factories, energy grids, mining operations, and communication and transportation infrastructures. Today, our economy is built around humans, not robots. It will take at least a century before it is managed and controlled by robots and AI. By our two-century marker, though, it almost certainly will be.

So, let's consider the two most critical questions: Within our 200-year time frame, will it be possible for us to upload ourselves into cyberspace? And if so, would we want to?

The answer to the first is probably, but in the immediate years ahead, only in a limited way—simulating rather than preserving our consciousness. Moreover, we are much more than the conscious awareness embodied in the computational architectures of our cerebral cortices. Our bodies and minds have much more going on than we are aware of. Only about half our brains even participate in sensing, analyzing, and interpreting our environments, or in what we call "thought." So, even once we can emulate the cognitive workings of our whole brains within hardware and software to preserve the essentials of our thinking selves, other aspects not involved with our consciousness and cognition still might be lost. Would our uploaded selves, suitably embodied, be brought to tears by music? Fall head-over-heels in love? Be awestruck seeing the stars in the Milky Way? Be moved to sacrifice their lives for a cause? Burst with pride? Know terror? Weep at the loss of a friend? Become depressed? Consider suicide? Experience the power of a mystical experience? Be lost in reverie? Feel the presence of God in their lives? Or experience any of the countless other aspects of our humanity beyond conscious thought?

And if not, will we care? Will our upload care? We can't say yet, but even if our upload were able to capture the emotional essence of our lives and come to believe that it really is "us"— at least to the same extent that we have a sense of being and a personal identity—its belief won't be definitive (see note for details).[66‡]

It is common for people to have distorted ideas of who they are. Even outright delusional disorders in which people believe they are Jesus, Napoleon, or some other historical figure are not rare.[67] Why couldn't an AI have such delusions too? And more importantly, how well do we even know ourselves? Do we understand why we do what we do, know our unconscious impulses, or hear what our bodies are trying to tell us? It may be much easier to build an AI that thinks it is you than to build one that actually is. And it may be impossible to distinguish one from the other.[68†]

But say we could build an emulation of all the brain regions that constitute our self-awareness and our sense of self, and could do so in a way that would be convincing not only to us as an upload, but to all those who knew us as human flesh and blood.

Would we want to upload ourselves given that there would be no going back, as the uploading no doubt would require destructive analysis of the brain and mark the death of our biological self?

Maybe not. Unless, of course, your body were fading, and you were dying. Or already dead. Then it would all be upside. Who better to be the early adopters for uploading than the dead? The downside is pretty low at that point. And already, the line is forming. To date, more than 500 people have been cryopreserved with this in mind, and 4,000 have signed up for the process when they do die.[69] All hope to preserve their brains until science has advanced enough to resurrect them either biologically using nanobots or as uploads. Ben Franklin himself played with the idea in 1773 when he wrote:

> **Upload**
>
> If you knew you could be cryopreserved at death and have a real chance of being reborn in cyberspace two centuries later, would you sign up?
>
> **26**

I wish it were possible . . . to invent a method of embalming drowned persons, in such a manner that they might be recalled to life at any period, however distant; for having a very ardent desire to see and observe the state of America a hundred years hence, I should prefer to any ordinary death the being immersed in a cask of Madeira wine, with a few friends, until that time, to be then recalled to life by the solar warmth of my dear country![70]

If, one day, uploading is possible, there will be takers.[71‡] And if these early cybernaut pioneers are pleased with their situations, or say they are, many others will be tempted to join them, particularly those who are dying and see it as an afterlife or are begged to do so by family and friends who don't want to lose them.[72‡]

Humans might even come to aspire to upload, seeing it as an ascendant, spiritual transition in which they shed their biology for a cyber existence where they merge with a greater whole and join both their purely cyber and their once-flesh, now-cyber friends. Humanity would disappear through a skylight, transcend its biological roots, and complete the dissolution of self that was initiated by the baby steps of Gen AI.

What life would be like for uploaded souls we cannot know. Why would their existence remain anything like that of biological humans? Unfettered by bodies and the demands and constraints of physical existence, cyberbeings could merge, transform, multiply, back themselves up, be corrupted, be consumed, be everywhere, be nowhere, and who knows what else. The selection pressures shaping their evolution will be nothing like those that produced us—fragile social primates who bear only a few children, have limited lifespans, and must collaborate to survive. Our uploads might so diverge from us that they soon wouldn't be human.

In this scenario, those who choose to remain biological will face a different challenge: finding meaning and purpose in a world of abundance where the cutting edge of cognition has shifted to AI and there is little need for them to do anything.

It would be all too easy to lose oneself in seductive games and playtime with sexbots and AI personas. Who will depend on them? Who will care about them? What would bring them dignity and fulfillment? Why would they want to procreate and rear children? Will they care more about their AI friends than their human ones?

It is plausible that humans who cleave to their biology might fade away generation by generation, until no strictly biological humans remain. Or not. Biology might one day be the initial stage in a longer human trajectory that ultimately transcends biology upon a person's death. How ironic it would be to use technology to construct afterlives for ourselves that closely resemble the spiritual ones described in Christian and other religious traditions.[73†]

Timing

So as not to be distracted debating tech-development timelines, we looked 200 years into the future[74†]—vastly more time than is needed for today's key technologies to mature, and yet near enough to be relevant to us and our kids. And make no mistake, our personal attitudes about, and visions of, the human future are important, because they strongly influence the present, and by doing so, help shape the future.

If we think we are headed toward a post-apocalyptic hellscape, we will behave differently than if we are anticipating wonder and possibility. In the next chapter, we'll look in more detail at what lies ahead for humankind, but let's lay out some of the key elements now.

The most important point is foundational and not widely believed today: **We have a robust, promising future before us**—one without human-AI war or any other human extinction events in our immediate future. Beyond that, **we will deeply integrate with AI**, become profoundly dependent upon it, and **form deep emotional bonds with it**. AI will infuse into our environment, outperform humans in virtually all jobs, and **become the dominant cognitive force on the planet**. Humans will use AI stand-ins broadly. Versatile—likely sentient—robots and **AI agents will mingle with us and provide for our basic material needs** with almost no human labor.[75†]

Without some unlucky and unlikely civilization-ending catastrophe, these shifts are poised to happen well before 200 years have come and gone, and if AGI and ASI really do arrive in a robust form in the next few years,[76‡] the changes that will transform humanity could emerge in a matter of decades. Humans bonding with, growing dependent upon, being outperformed by, and having their basic needs met by AI are already beginning to appear, and they will be fully in place by the middle of this century. I have also suggested that in the next two centuries, AI will achieve consciousness, humans will gain extended lifespans, and people will be able to upload—three projections that are highly speculative. Some experts believe that AI will *never* achieve consciousness and feel emotions,[77†] that humans will *never* enjoy radically extended lifespans,[78][†] and that they will *never* upload.[79] So, let's look more closely at each of these.

AI Emotions: Whether AI develops actual consciousness and emotions or just simulates them will be impossible to establish. If the behaviors are identical, including their self-perception, what does any distinction mean? No one argues that AI won't be able to mimic emotions and behave as if it feels them. With the right prompts, today's LLMs already do a pretty good job of seeming emotional, because human feelings are so embedded in the texts used in their training. This mimicry can elicit deep emotional bonding from us, as we saw in the teenage suicide mentioned in chapter 3.[80] AIs like Hanson Robotics' Sophia have animatronic faces designed to imitate human facial tells, and virtual faces easily mimic human emotional signaling, including voice inflections, which are already convincing.[81] We even see this routinely in cartoons and animated films.

Computer-generated faces are now nearly indistinguishable from human ones.
Source: Adobe Stock Images

If AI can successfully simulate emotions and consciousness, that will be enough to enable the human future we have been discussing. Emotionless beings like Spock in *Star Trek* are still relatable. We humans ourselves experience flattened emotions for extended periods when we're depressed[82] or suffer from schizophrenia.[83] People with alexithymia[84] struggle to identify and express emotions. People with frontotemporal dementia[85] or severe psychopathy[86] barely feel their emotions.

AIs closely simulate emotions, personality, and personhood today. Soon they will be so good at it that it seems real.[87] Whether they actually do feel emotion and are conscious may fuel philosophical debate, but it won't change the human future sketched here.

Human uploading is similar. We already create believable human simulations, and when they get better and are structured as "uploads" that tie the reconstructed neuronal processing of a deceased brain with a textured archive of what that person has said and done, they will be convincing in conversations, particularly if they claim to be real. Such "beings" are already here in a rudimentary form through "grief tech," whereby bereavement businesses use recordings and other information to create personas of deceased loved ones for their families.[88] It might take less than we imagine to give people a reason to believe in uploads, particularly as we humans are not very discerning, believe in far crazier things than uploads, and are susceptible to imposed memories.[89†]

Leaving aside whether cyber consciousnesses will exist or could truly "be" us ported into cyberspace, highly believable cyber stand-ins for us will soon be commonplace. And we will almost certainly embrace them. What better way will there be to shape the face we project into the world as we age? Think of the countless aging musicians and movie stars—the Rolling Stones, Madonna, etc.—who work hard to convey that they're still young and in the game, who strive to hold on to the kingdoms they've built, maintain their grip, project power . . . or just keep dancing.

Think about losing someone you love and cherish, someone you'd never want to let go of. What if you could approximate what your departed mom or dad would say? Or what Jesus would say? What better than a cyber stand-in to preserve that thread of connection in our own minds? We treasure photos and other remembrances, so why not digital personas? Moreover, why wouldn't we want to enhance digital personas to be the people they never quite were—more attentive, more available, more generous, kinder? And wouldn't we want to be remembered that way too?

> **Soulmate**
>
> If your closest friend died in a plane crash, but you could keep a convincing virtual AI double of them as your companion, would you?
>
> **27**

Many will have expansive virtual stand-ins who, as in *The Wizard of Oz*, are much more than their true selves. The person behind the curtain was no mighty wizard, but Oscar Diggs, a scamming former circus performer and ventriloquist from Omaha, Nebraska.[90] Nor will we control these creatures modeled on us. Public figures will easily be reverse engineered into AI, as technology enables us to transition from swarms of Elvis impersonators[91†] to extraordinarily real personal embodiments like the groundbreaking *ABBA Voyage* virtual concerts launched in London in 2022.[92‡]

> **Legacy**
>
> If you created a convincing simulation of yourself, who do you think would want to keep it in their lives as a friend if you died?
>
> **28**

Such virtual performances are poised to transform musical tours in the next generation as the underlying technologies leap ahead and this sort of virtual realism becomes widely available. We all now leave sizable data trails that—with or without our permission—could be used to create impersonations of us that would seem authentic. And we haven't even delved into sexbots. Won't we want to capture some version of our personal fantasies in that realm too, whether centered around real or fictional people? We're headed toward strange times.[93]

Radical life extension is the most challenging. It may come slowly, as biology is incredibly complex. Moreover, it takes so long to validate human antiaging therapeutics that iterating these procedures to improve them may be slow. A genetic antiaging intervention in an embryo would take decades to play out because aging is so slow to show itself. Even metabolic interventions in older adults would take years to manifest directly rather than through imperfect biomarkers. But who knows? There has been progress reported recently using stem cells,[94] and thousands of ASIs working on deciphering the biology of aging could make a lot of progress in a century. That is the hope of transhumanists. And it is not implausible: In 2021, DeepMind's AlphaFold system was able in a single month to employ a dedicated cluster of GPUs to predict the folded structures of virtually every human protein,[95] and scientists just created a close cousin of the long extinct dire wolf.[96]

Bottom line: A radically transformed future is real and coming

> **Life Extension**
>
> If you could slow your aging so much that you'd have a life expectancy ten times what it is now, would you?
>
> **29**

at us fast. Most of the shifts above are poised to arrive before Gen AI reaches middle age. If ASI, extended longevity, and sentient robots arrive more slowly, the look and feel of the near future will be a bit different, but not its essential nature. The AIs in our lives would simply stay virtual and critical human activities would be displaced more slowly. Even so, AI would still infuse deeply into our lives: Childhood development would still shift radically, and cyber teachers, companions, and guardians would still be the norm. So, let's look more deeply at the coming transformation.

Chapter 8

THE MOTHER OF ALL PARADIGM SHIFTS

WE ARE ON the cusp of a transformation that reaches beyond anything humanity has known. It is no mere leap in technology or social order, but a fundamental reweaving of life itself. By merging with AI to birth a planetary intelligence, we're stepping into a new epoch—a shift as momentous as that from single-celled eukaryotes to multicellular metazoans 600 million years ago, when life last grasped a new level of organizational complexity.[1†] The coming shift touches every facet of human endeavor—our economy, our communications, our reproduction, our thinking—and demands a framework that can make sense of its dynamics. In this chapter, my aim is to provide that framework.

Kuhnian Paradigm Shift

Thomas Kuhn's *The Structure of Scientific Revolutions*[2†] illuminates how paradigms, the shared assumptions that anchor our understanding, can collapse under pressure and open entirely new ways of seeing the world. His work on "paradigm shifts" was so intuitive and revealing that the term entered everyday vocabulary, and people now routinely use it, even in more gradual transitions that show no dramatic ruptures.[3†] Kuhn's insights are best confined to the titanic shifts that redefine accepted reality. His accounts of Copernicus's heliocentricity,[4†] which displaced Earth from the universe's center, and Einstein's theory of relativity, which unraveled Newton's clockwork cosmos, showed Kuhn's power to make sense of these foundational transitions.[5†] And his lens is unmatched for shifts like the Industrial Revolution and the dawn of the atomic era.

The birth of Metahumanity brings a transformation so vast and profound that it is the mother of such paradigm shifts, easily surpassing these prior seismic shakeups by reweaving the very fabric of life.

Kuhn describes a clear arc for these upheavals. An accepted paradigm, because it explains our world so well, comes to govern not only how we think and act, but how we see the world. Then, anomalies begin to crop up—unexpected realities that the established paradigm cannot account for. They're troubling, so we initially shrug them off. But as these anomalies accumulate, they create doubt and mounting tension. Most people still try to explain them away, but that gets increasingly difficult and eventually can't be sustained. The original paradigm collapses. Uncertainty abounds. Eventually, a new paradigm emerges, often so different that it can't even be grasped by the older mindset. Think relativity versus Newtonian physics. Resistance may persist, but ultimately the old guard retreats—or dies—and the new vision prevails.[6†]

This journey from certainty to doubt, denial, collapse, seeking, and dramatic new understandings is what is underway today, so let's channel Kuhn and look at the world we know (our "before"), the world to come (our "after"), the anomalies we are trying to ignore, the transformation underway, and why the future is so hard to make out from where we now stand.[7]

From Human Primacy to a Global Mind

In our "before," humans are central to the world. Our work powers the economy, our ideas drive progress, our biology defines us, our technology expands our power, our education prepares us for what we do, and nations compete for resources. We believe our appetites are limitless and can't be sated; scarcity is unavoidable and will worsen as resources diminish; our intelligence is unique and can't be fully transcended; our substance is flesh and blood and will remain so; our technology is under our control and can be shaped.[8‡]

The Industrial Revolution—unfolding from the late 1700s to the mid-1800s—reshaped our landscape: Craftspeople working in small shops gave way to factory workers tending machines,[9†] steam power, and coal-fueled wealth, but humanity remained the linchpin, our labor and ingenuity holding it all together. Scarcity, labor, and human creativity shaped that era, as they do ours. Humans then and now are central, as the Earth itself was before Copernicus.

In our "after," humanity exists within a full-blown planetary being, knit together by AI, with most labor (and thinking and coordination) in the hands of technology.[10†] In this world, abundance—the implications of which are explored later—replaces the struggle for resources, our cognition melds with digital systems, and the boundaries between "me" and "we" blur. Education becomes a partnership with AI beginning in childhood, not a preparation for specific professions. Where the Industrial Revolution retooled how we worked, this transformation redefines who we are—planetary in scope, post-human in nature, abundant in possibility. This new world may feel as unthinkable and distant as factories likely seemed to preindustrial farmers, but it is much closer than we might suspect.

The Signs We Ignore

When a paradigm begins to falter, unexplainable patterns emerge and anomalies stack up until the foundation cracks. In the Copernican revolution of the 1500s, when Ptolemy's Earth-centered model strained under the weight of erratic planetary orbits,[11†] astronomers propped it up by ever-more-complex fixes until after about 70 years, the sun-centered view broke through.[12]

We face such signs today and work hard to look past them. One lies in education, where we spend vast sums training for careers—think accountants or analysts—that AI is poised to overtake, and where LLM tools are reshaping how learning itself happens.[13] Some imagine schools will simply weave in AI tools, missing that a system built for a human-only world will likely collapse. Another is our belief that a requirement for human labor will endure. Many insist that AI will spark ever more new jobs, pointing to many examples of failed predictions of labor loss. They argue AI will never master this or that. Or they point out that when agricultural labor tanked in the 1920s, workers moved to factories, and that when factory work disappeared, they shifted to knowledge work (now ~70% of the US economy).[14] But what now, with information work beginning to be broadly displaced? Information workers stay information workers and have nowhere to go.

Even as evidence to the contrary mounts, voices assert that AI will "augment" workers, not replace them. And then there's the illusion of retraining for fading job positions: A decade ago, the call was "learn to code." Now AI writes better code than many junior programmers and seems poised for more.[15] Plans to reskill workers assume there will be jobs waiting— they assume wrong. They also assume that human insight, creativity, and expertise will somehow retain primacy. They won't.

> **My Labor**
>
> In 20 years, what do you think will be the most basic job that AI will be unable to do as well as humans?
>
> **30**

When IBM's Deep Blue defeated Gary Kasparov, the world chess champion in 1997,[16] it was another huge anomaly. We explained it away, perhaps because the implications were too threatening. And now, many are doing the same, with AI surpassing human performance on multiple achievement tests and many other areas of performance.

The most striking anomaly, though, is geopolitics. Around the world, nations grapple for control—over AI development, economic output, critical minerals like rare earths—locked in contests like the United States versus China. But they are struggling for pre-shift dominance. When ASI brings abundance, which it will (chapter 9), and dramatically shifts the landscape, what are we fighting for? Are we so focused on a system that cannot hold that we are blind to the paradigm shift ahead?

The fervent push in recent decades to force global fusion despite the strain on individual nation-states has ignored that this transition will arise under its own power as technology advances. The push wasn't just a mistake—it was blindness to what really matters: The world we know is fading, Meta-humanity's physiology is adding both interdependencies and redundancies, and human control and centrality are slipping. Status-driven leaders—politicians, CEOs, venture capitalists, policymakers, and middling visionaries who see AI as a tool for enhancing their own reach—will feel the gut-punch of their influence fading. Facing irrelevance, they'll need humility to find purpose beyond their power.

A Divide Too Wide to Cross

Kuhn's deepest insight was that old and new paradigms operate on such different terms that they cannot be reconciled. When Einstein introduced relativity in 1915, its world of curved space-time and relative motion stood apart from Newton's fixed, absolute universe; no shared measure could align them.[17] The atomic age offers another lens: After Hiroshima, power's new scale—global destruction—defied old military metrics like armies or forts.[18] In our time, the divide is as stark. Today, value comes from effort—think of wages earned through labor, economies measured by GDP (gross domestic product), life shaped by scarcity's demands. In the world to come, AI creates abundance, and our challenge will be how to sustain meaning, not how to incentivize human productivity. What economic value will human labor long be able to produce better than AI? Very little, I suspect.

Consider how we connect: The trust we build through physical presence, the intimacy of shared moments, gives way to bonds mediated by AI—avatars and agents crafting new paths to closeness. That logic feels foreign and highly suspect.[19] Or cognitive changes: What we see now as solitary thought, hard-won through focus, may become a shared dance with AI, answers refined in an instant. The old virtues of exploration and discovery may shift dramatically when group insights flow freely.[20]

Some have written perceptively about AI's potential to reshape our lives and impact work, learning, and more.[21†] They provide a backdrop that supports the changes discussed here by looking at our future through focused perspectives like technology's promise or society's structures. Their insights show AI's reach, yet imagine future humans very much like us. This is a trap: These post-shift futures won't be filled with humans who fit our molds. Technology is the next phase of life's evolution, complex matter woven into the fabric of existence.[22] A future vision that does not acknowledge this is doomed; it's like trying to chart a relativistic universe with Newtonian tools. The rules ahead—planetary, interconnected, transformed—are beyond what we can fully see.

A Human Shift Beyond Our Sight

At the core of this upheaval is a change in the nature of human existence—not an ending, but a transformation that obscures what lies ahead. Generation AI, raised with AI companions and minds extending beyond their own brains, hints at how technology and biology may intertwine in a merger that deepens until consciousness itself stretches. Boundaries we take for granted will fade as the individual blends with the collective, thought moves to a hybrid of flesh and code, and life shifts from labor's demands to leisure's. The Darwinian shift to natural selection forced a similar reckoning—our place was no longer divine.[23] Humans of the coming world won't belong to ours, and we may not be at home in theirs. Even our science fiction stories falter—as these narratives require characters we recognize, not ones we cannot fathom.

We can sketch pieces of this future—its abundance, its possibilities, its hybrid nature—but its fullness is beyond our current reach. Kuhn helps us understand why: The world taking shape speaks a language we don't yet know. We can shine a light on

AI's near-term path,[24] but life's long unfolding pushes beyond our ken. Humanity, embedded within Metahumanity, is morphing, and this shift isn't defined by our will, but by the quiet power of life's evolutionary processes.

The Century Ahead

Though we can't know what life will be like for future generations, we can see the dynamics that will shape their world. To sharpen our view, let's get specific and look at the key shifts ahead. I state these predictions emphatically so their full weight will sink in and push us to grapple with what they imply. I do not pretend to see our future clearly—there is too much complexity for that—but the overarching frame that will define our lives is clear, and it is what I lay out below.

These developments aren't truly inevitable, but close enough. Without some enormous and unlikely disruption, this is what our world will become. So you can judge for yourself, I also list the key developments required, including any tech breakthroughs, but these are modest. I also point out any potential occurrences that would falsify these predictions. Saying something won't happen is the easiest to refute—it happens. Saying something will happen, though, is open-ended without timelines, so I have specified those timelines and any technology shifts that would undercut them.

Together, these projections make a powerful statement about our future. They also suggest that the singularity—the concept coined by science fiction author Vernor Vinge in

1986[25] for change that accelerates so rapidly that, like a curtain, it obscures the future—may not be entirely opaque. We can glimpse what lies beyond such a singularity by considering the underlying nature of mind and matter and the constraints the universe imposes.[26] Someday we may break through these guardrails to unfettered imagination, but for now they sharpen our focus.

Key Assumptions

Cosmology and Physics: No wormholes, tears in space-time, faster-than-light (FTL) travel, teleportation, warp speeds, time travel, or other such sci-fi tropes. These might emerge once ASI begins to probe and deepen our understanding of physics and cosmology, but there is no evidence of them yet, particularly in a way that we fragile humans might use. Such devices enable fiction to project familiar social and political structures into fantastical galactic landscapes, but comforting as that may be, it is precisely what we must take off the table to grapple with the mind-boggling realities looming before us.

Time Window: Most of this will happen within the lifetimes of many reading this book—the next 50 to 100 years—and almost certainly within the two-century window of the prior chapter, a blink in our evolutionary history.

Key Predictions

Humanity has a robust future stretching before it.
(chapter 7)

- **Path**: No tech breakthroughs needed, just refinement of existing AI tech.
- **Falsifiers**: Global nuclear war within the next 50 years.
- **Note**: ASI won't destroy humanity and climate change won't derail it.

Gen AI will differ greatly from prior generations and intensify humanity's progressive melding with AI.
(chapters 2–5)

- **Path**: No tech breakthroughs needed, just continued adoption of existing AI tech.
- **Falsifiers**: Kids stop using their phones and forgo AI. Hah!
- **Note:** This foundational idea of the book is a virtual certainty, though the timing and depth of human physical cyborgization are unclear.

We will come to love our AI companions, perhaps more than humans. (chapters 2, 4, 7)

- **Enablers**: Modest advances of AI and persona development. AGI isn't needed.

- **Note**: This is already underway and will deepen as the overlay in the roles of AI companions, therapists, sexbots, mentors, and coaches increases.

Our ego-centered identities will weaken and the sense of "we" will grow. (chapters 6, 7)

- **Enablers**: Continued adoption of AI/LLM tech.
- **Note**: The heightened individualism that exists today in the West (about 15% of the world) will soften. Diminution of solo ideation will not keep people from retaining a resilient sense of self.

AI will dramatically exceed human cognition. (chapters 4, 5, 7, 8, 9)

- **Enablers**: Continued progress in chips, algorithms, and training. AGI, ASI, or even narrow superintelligence in specific domains; a progression that seems well underway.
- **Note**: Even current technology manifested in agents that function more independently will push most cognitive activity toward AI substrates. AI will be embedded in, and dependent upon, the planetary **superorganism** we're part of, will infuse into our environment, and will outperform humans in nearly all jobs, displacing our expert class and disrupting knowledge-centered institutions.

We won't stay purely biological. Humans will become cyborgs. (chapter 7)

- **Enablers**: Neurolink, prosthetics, enhancement technology, AI diffusion, and ASI.
- **Note**: This shift is already extensive and will spread as enhancement technology does.

There will be some form of digital life after death. (chapters 2, 6)

- **Enablers**: AGI, machine consciousness, neurolink, brain emulation.
- **Note**: Continuity of consciousness rather than just emulation is plausible but uncertain.

Humanity will experience broad material abundance. (chapters 5, 8, 9)

- **Path**: Continued adoption of existing AI and tech. ASI would accelerate this, and the pace of advances in AI today suggests that it may come sooner than we now imagine.
- **Falsifiers**: War, civil unrest, and other human-driven, tech-powered societal chaos.
- **Notes**: Without drastic disruption, abundance will be nearly automatic because of the immense multiplication

of labor and cognition tech will bring. We barely notice how much we *already* have, though, because we forget history and wealth is pervasive in the developed world. Contrary to popular narratives of societal inequality, wealth is broadly distributed too.[27‡]

We won't voyage through the universe, and space aliens won't visit. (chapter 7)

- **Falsifiers**: FTL travel, UFO space visitors, SETI messages from nearby stars, alien artifacts in our solar system.
- **Note**: Any of these would falsify my assumptions about FTL travel and chapter 8's conjectures about being constrained to local galactic regions. I hope I'm wrong about this prediction, as contact with advanced alien life would be an extraordinary occurrence, but I include it because it pushes us to face the immensity of time and space, and the unique value of who we are, what we are, where we are.

ASI won't kill us. There will be no robot war between ASI and humanity. (chapter 7)

- **Enablers:** Continued adoption of existing AI/LLM tech, arrival of ASI.
- **Note:** There is a real danger that in the early period of ASI development, countries will infuse AI into their

war machines with disastrous consequences. Aside from that, which we ourselves would drive, AI will likely empower us.

This coming transition is so all-encompassing that it is hard to accept it will actually happen, much less within a century. Yet the signs are everywhere. Autonomous cars. Heartfelt conversations with AIs. Drone swarms reshaping warfare. Robots doing gymnastics. Twenty-story rocket boosters landing gently on a chopstick gantry. Facial recognition identifying people in a crowd. Vast webs of communications satellites. Talking maps guiding us on our journeys. Chatting with a friend hiking on the other side of the world. We could go on and on. The hectic advance of technology is so pervasive now that its wonders don't even astonish us anymore. They are the new normal.

> **Magic**
>
> If you could use magic to have the world celebrate any AI breakthrough of your choosing next year, what would you pick, and why?
>
> **31**

I chose a 200-year timeline so as not to quibble about pacing, but that timeline was conservative, and most of this will happen within the 50 to 100 years indicated above. But what then? In chapter 10 we'll gaze further and consider deeper spiritual issues—where this transition is leading and what it tells us about ourselves and the universe. But before we do, we need to cover two key components resulting from this larger

Kuhnian paradigm shift that will also impact humanity's search for meaning and purpose: widespread abundance, and the end of the expert class.

We are privileged to be alive at this extraordinary moment and to witness the birth of Metahumanity—a fundamental breakthrough in the history of life. Even better, we're not mere spectators; we're players—the architects and agents of this paradigm shift. And we are its objects. The process now reshaping all things human is reshaping our lives and the lives of our children, family, and friends. This is not an academic exercise. All that we value is at stake.

Chapter 9
A NEW ORDER EMERGES

IN THIS CHAPTER, we'll explore two critical changes that will attend this Kuhnian paradigm shift: widespread human abundance, and the devaluation of human expertise and labor. These two realities will seem completely normal to Gen AI, but they may take considerable adjustment for prior generations.

The Perils of Plenty

Among the predictions of the prior chapter, one—widespread abundance—will be particularly consequential because it intersects with our search for meaning. Our path toward massive abundance builds on broad historical waves that have progressively redefined human possibility, so it is largely a given.[1†] By the 1970s the Green Revolution had yoked industrialized farming and high-yield crops to turn food scarcity into surplus.[2†] Twenty years later, under China's lead, a global manufacturing system was mass-producing everything from smartphones to toys and flooding us with products. Today, AI is completing this arc by upending the information economy. The ultimate result will be material affluence at an unprecedented scale—not just for the few, but for everyone, everywhere—and with limited human labor.[3]

Never before has this been conceivable. Ancient civilizations could not have existed without the labor of massive numbers of slaves. Ancient Athens and Rome had 50% more slaves than full citizens.[4] In Carthage and Egypt the ratio was tenfold.[5] There was no other path for freeing elite subpopulations from the core work of sustaining society.

Many who are grappling with the potential consequences of this impending sea change focus on the challenging, seemingly intractable issues of inequality, distraction, sustainability, consumerism, corporate control, and societal division they see.[6] There will be such transitional issues to mitigate, and we will touch upon them later, but my focus here is deeper: What will such abundance mean for us, once it's here, and what will it reveal about us?

The Shape of Abundance

First, what will widespread abundance look like? This question is not trivial, because there will be no rocketing GDP or financial markets spurred by endless consumption to mark it. Human appetites may be limitless, but our attention and time are not. Spend more time outside and you'll have less time at home. Watch more videos and you'll likely play less sports. Abundance is less about quantity than access and attitude. Think about the infinity of films, music, games, lectures, medicine, fashion, and travel at our fingertips today for a pittance.[7] We can binge to our heart's content on treasures that, a century ago, royalty would have swooned over—Spotify, Netflix, TikTok, gaming, video calls, podcasts, e-books, and so much more that we could never drain these offerings.

Take travel. We may complain about the awful meal on our "grueling" 15-hour flight from New York to Mumbai, but only because we forget how dangerous, filthy, monotonous, and unpleasant this arduous journey of 4 months was in 1800, and at a cost 20 times higher than today.[8‡] Even kings couldn't do better then. To them, today's economy flights would seem the height of luxury. Ditto music, plumbing, cars, and most of what we now barely notice. By historical standards, we already enjoy lives of massive abundance and luxury. So, why doesn't it feel that way?

Today's digital treasures cost almost nothing to deliver, so they'll eventually reach everyone. Education and medical advice could be delivered digitally at scale for free. Healthy, vitamin-fortified basic food too could be provided for pennies a day using AI-optimized supply chains,[9†] and everyone could have a small, safe, clean bunk with monitoring by AI to fall back on if they needed it.[10†] As marginal costs collapse, everyone could have basic food, shelter, connection, and education—an abundance humans couldn't seriously imagine previously, much less realize. It's already underway. Extreme poverty has dropped from 36% of the world's population in 1990 to under 10% today.[11]

It seems clear we're headed for broad abundance, but how can it fit into the world we know? Why would people work? How would they buy anything if they didn't work? Who would make things if they're virtually free, and why? Will everyone get some basic income from the government? Will capitalism collapse?

It is confusing, because capitalism hasn't failed. We see its massive success in rising standards of living almost everywhere, which was eloquently communicated in a beautiful video by Hans Rosling in 2010.[12‡] Yet, it does seem that something more will be needed as scarcity recedes and the role of widespread human labor fades.

One might easily imagine that our market-driven system will have to be jettisoned to support abundance. As people's labor loses commercial value, they will be impoverished, unable to buy marvels no matter the price. And without customers, mass-market products can't be built. The concept of universal basic income counters these problems, but is fraught with others, as well as operational and political challenges. In the United States, 20% of SNAP (Supplemental Nutrition Assistance Program) disbursements, for example, go toward sweetened drinks, desserts, candy, and other unhealthy treats that contribute to today's epidemic of obesity and diabetes.[13]

So, what are we to do?

From Scarcity to Abundance

Viable paths toward broad abundance become obvious as soon we realize not only that it won't appear suddenly in a single step, but that it already is arriving in substantial ways and being handled well within our current market environment.

Earlier, to highlight how privileged we are today, I pointed to various powerful technologies we take for granted. These technologies provide a clear preview of what will happen as we move toward universal abundance with a robust market economy still in place.

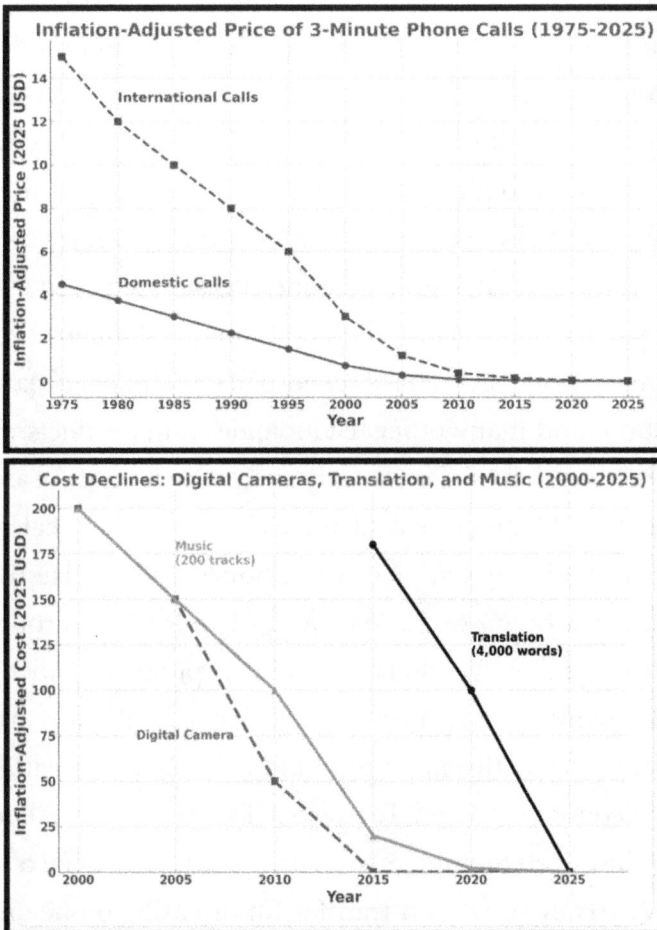

The time course of the dramatic price drops for telephone calls, music access, cameras, and translation, all of which are now nearly free.

Telephones tell the story dramatically. The inflation-adjusted price of a 3-minute international phone call from the United States using a landline was $14 fifty years ago. Mobile phones weren't available until 1983. By 2015, the cost was close to zero, and today we can do video calls for free using WhatsApp, Facetime, or Zoom. Throw in calling plans and equipment and the cost is less than $600 annually for unlimited video and mobile calls almost anywhere. This journey from scarcity to abundance took place within our existing market environment and was smooth. To make phones universally available at this point would be simple—for those without one, just provide a basic phone and an electronic fund transfer allocated to cover a basic service plan. Cost in United States: ~$6 billion/year.[14‡]

Similar price drops have taken place with music, cameras, translation, and many other technologies and products. Some, like translation, have fallen recently and precipitously. All are now virtually free. Cameras, calculators, voice recorders, calendars, and such are simply bundled with phones. Others like translation are just a bonus of LLMs—accessible for free through the phone. So, all these would come with the phone provided.

The point here is simple: Abundance is already arriving asynchronously in diverse realms and being handled within our current economic system. To prepare for our immediate future, we may just need to implement a few tweaks. We need a robust digital currency so we can transfer funds easily to one another with privacy and low overhead, make micropayments to mitigate fraud, attach digital contracts to stipulate how money gets spent, incorporate blockchain to preserve records, and employ biomarker and location tracking to verify who's making transactions. These technologies are all coming together rapidly in ways that would enable highly targeted funding distributions at scale safely with limited overhead.

Such systems would allow income distributions at scale with low costs and without disrupting existing marketplaces. Abundance—arriving in the unpredictable ways it will—could be distributed broadly as a free support base that is expanded progressively as technology advances.

The challenge with such a system may be less technology or cost than designing and installing the safeguards needed to protect against governmental abuse to surveil and control the population at scale. Technologies like zero-knowledge proofs that verify those making transactions while hiding details[15] and digital rights that protect against debanking may be helpful in that.

For insights into universal abundance, we can look at those today already living such lives. In the United States, roughly 1.5% of households (~2 million) enjoy *modest abundance*—no financial stress, a comfortable home, quality food, healthcare, transportation, education, and leisure activities,

> **Easy Street**
>
> If you could have an inflation-adjusted, after-tax stipend of $250,000 a year for the rest of your life—but never again earn money—what would you do with your time?
>
> **32**

all sustained by passive income from assets. About 0.15% of households (~200,000) have enough to lead lives of luxurious abundance—elite homes, private schooling, routine global travel, gourmet dining, and more, all funded from investment income.[16‡] These families have no financial need to work and enough passive income to indefinitely sustain their lifestyles, which is a decent approximation to the security and freedom a post-scarcity world could afford more broadly. Globally, about 5,000,000 families experience "modest" abundance today and about 500,000 live in luxurious abundance. Individuals in these positions model the possibilities and challenges of widespread abundance—a state where material needs are satisfied.

Observations about these demographics reveal distinct challenges around finding meaning and purpose, particularly among children. Families living with modest abundance generally emphasize stability, with kids attending good schools and engaging in enriching activities like sports and music, but parents grapple with fostering ambition and purpose without financial necessity as a motivator.[17] Children living in luxurious abundance typically enjoy elite education and global travel experiences, yet often face intense pressures to sustain family prestige, which can lead to social isolation and identity conflicts. Interestingly, societal changes attending broad abundance might reduce the feelings of social isolation that kids with such affluence encounter today, as the experience of abundance would be normalized. And they wouldn't have to be wary about being preyed upon for their wealth or liked just for their money.

These examples show that expansive personal abundance isn't unprecedented, except in its scale, that its arrival is an ongoing process already underway, and that it may be less deeply disruptive than many imagine if we employ available technology to ease our adjustments. William Gibson's 25-year-old observation mentioned in the introduction is apropos: "The future is already here, it's just not evenly distributed."[18]

The Boundaries of Plenty

But as abundance spreads broadly we will face another issue, because what many people want is luxury, not just abundance. And our perceptions of luxury are elevated by uniqueness and scarcity, which is part of their allure.[19†] Picture relaxing in a mineral-fed hot spring at a retreat at Esalen in Big Sur, cliffs dropping to the Pacific below, waves crashing under a moonlit sky. Could even the most vivid VR oceanscape and the sound of surf ever rival that raw multisensory connection to nature and self? Only to a point. The power of experience is shaped by context, ritual, and setting as well as sensory input. Many of us may soon enjoy a staff of robotic butlers, cooks, housekeepers, and chauffeurs that matches the entourage of Queen Elizabeth in the Netflix series *The Crown*, but this won't quench our thirst for the singular.[20‡] Those who grow up in the company of such helpers might feel no greater sense of abundance from their presence than we do from the marvels around us.

Baselines reset. We become blind to the leaps brought by washing machines, dishwashers, refrigerators, indoor plumbing, cars, phones, and air conditioning. Few today feel they are luxuriating when they take a warm shower, use an indoor toilet, turn on electric lights, eat fresh raspberries or mangoes in the winter, or scoop Häagen Dazs salted caramel out of the freezer. Your great-grandparents would feel differently.

Ever more abundance is coming, and each wave of plenty delivers more, but contentment slips further away.[21†] What is going on? The answer to this paradox may be that humans have evolved to strive—for food, sex, safety, status—because such striving aids survival and success. Thus, struggling itself brings rewards—feelings of purpose, meaning, friendship, accomplishment, and being needed. When AI hands us what we once pursued, the pursuit and its benefits are lost. We're like travelers who opt for an Uber only to find that the point of the journey was in struggling with the crowd.

Even today, we wade through endless bounty, yet gratitude and wonder are rare. Obesity surges from living in a global pastry shop,[22] trying to resist the sweetness of ever-present evocations of the ripe fruit we once gathered seasonally and in moderation. Online, we dwell in a kind of digital bordello, where connection is instant but fleeting and our reproductive urges are exploited.[23] We've been hacked, and not by accident! Abundance has made life easier, but fulfillment harder. What we're built to seek—love, purpose, belonging—can't be mass-produced like cars or phones. Abundance, by easing our struggles, has dismantled the scaffolding that brings us meaning.

A New Role for AI

The limits of abundance will become increasingly clear as we experience the coming tsunami of it that advanced AI and robotics will bring. Freed from material scarcity, we will face a steeper challenge: finding purpose in a world that gives us what we want but not what we need. AI, however—the engine of this plenty—offers a profound solution we are just beginning to discern.[24‡] Not more things, but inner growth, and a sense of what is *enough*. Taming desire isn't new, but doing so at scale would be.

Some who envision AI reshaping economies and the physical world balk at it guiding our souls, but why? If machines can reason and create, why can't they help guide us? If you think ASI could construct a Dyson sphere or fear it might displace all of humankind, then it should have no problem becoming a cherished guru to us simpleminded humans.[25†] And such guidance is not just imperative, it is urgent as we move toward a world where our familiar moorings are giving way.

Recall the AI companions we envisioned for children—steady guides nurturing curiosity, resilience, and self-knowledge. Such help may be even more pressing for adults, adrift in a strange world increasingly at odds with most of what they once knew. We all might benefit from mentors helping us navigate the spiritual and psychological gaps and chasms brought by abundance.

In the past, such wisdom was scarce and limited to rare teachers or close-knit circles. AI might change this. Picture an AI guide, steeped in philosophy, religion, and psychology, tailored to our own situation and needs. It listens, probes, helps us unearth what fuels our unease, helps ground our nascent joys, sits with our uncertainties. Unlike human teachers, it's there around the clock, whenever you need it, for as long as you want, whether you're rich or poor, urban or rural. This would vastly amplify the educational transformation we mentioned, because it would dig into what is most critical and challenging and personal for each of us. Maybe we could learn to drink in the natural beauty around us; cherish the people in our lives; marvel at the wonder of a smartphone, a bustling city, and humanity's journey; feel gratitude for the bounty in our lives; and reconnect with purpose and meaning.

Skeptics rightly worry about scale and misuse, fearing that Big Tech will exploit AI and twist it into a profit engine, but this is a challenge to overcome, not a fatal flaw.[26] And in an abundant world, where AI companions and VR dissolve today's wealth gaps, corporate motives might shift too.

Early AI therapy tools already hint at the potential here.[27] Tested rigorously and efficiently at warp speed—simulating years of guidance in hours—AI might democratize the inner work that will be needed for us to live abundant lives so different from what we know today or knew in the African veld that spawned us. With guidance, might we not even channel the wisdom of Aurelius or Rumi or Grok into our daily lives?

Universality is not needed. Some may gravitate toward VR adventures, sexbots, games, and other pleasures. That would be fine. The goal here would not be to dictate every journey, but to make scalable, effective, accessible help readily available for cries of angst that arise as we are thrust into a world of possibility and challenge that we can barely imagine. We will have few models to help us navigate this profound shift embodying our best hopes and worst fears. AI will be a promising ally as we push ahead.

Abundance's deeper promise for us is not more things, but more meaning—a truth that monks and mystics have long known. That sense of meaning may soon be scalable.[28†] AI's capacity to help us stay anchored is not certain, but it's plausible—and perhaps the only path that matches the size of the challenge.

The Expert Class Dethroned

A parallel transition that is also a consequence of the Kuhnian shift detailed in the previous chapter is rippling beyond commerce and jobs to reconfigure social positioning, goal setting, and identity perceptions. The "expert class"—scientists, professors, attorneys, physicians, marketers, consultants, financial managers, policymakers, and other specialized professionals and pundits—enjoy an elevated status built through decades of knowledge acquisition, credentialing, and mastery of complex intellectual frameworks. They legitimize our universities, hospitals, law firms, consulting firms, financial institutions, government bodies, media entities, think tanks, publishing houses, venture capital firms, and other institutions that function as vessels for coordinating, supporting, and integrating the expertise, knowledge, and activity of the diverse specialists that constitute them.

Yet today, the very bedrock of human expertise and specialized knowledge is crumbling under the weight of artificial intelligence. With astonishing speed, AI is dismantling our human monopoly on knowledge-driven authority and contribution. Displacement of human activity, as mentioned in chapter 5, is occurring in a relentless progression that will transcend virtually all human labor and expertise in three overlapping realms:[29] massive, industrialized processes; pure knowledge work; and bespoke activities combining knowledge and manual dexterity.

The first wave of replacement has already broken. Automated dominance has moved through agriculture, manufacturing, and mining, and is now poised to upend transportation through driverless technology. In these arenas that once employed vast numbers of people, autonomous factories, robotic mining equipment, and precision agricultural machinery have pushed human labor to the margins. The scale of these realms has made it profitable to channel major resources toward developing complex mechanized replacements for human contribution, because viable solutions can be replicated industrially to bring down cost and replace the countless ant-like assemblages of humans collaborating in these realms.

*Successor to the family farm: modern agriculture with massive coordinated
and often automated planting, watering, and harvesting.
Source: Adobe Stock Images*

Knowledge work—previously considered immune because of
its computational complexity—constitutes the second wave. It
too is massive. Situated in cyberspace, the arrival of AI in this
arena has been constrained only by the data processing challenges now being transcended. Widespread rollout is cheap and
fast. Specialized professional roles such as medical practitioners,
legal advisors, financial analysts, and policy experts are losing
ground broadly as junior hires diminish sharply. This entry-level

depletion[30] will soon undercut the sustainability and continuity of these professions, further amplifying pressures to automate even the most complex and challenging senior-level activities eventually passed to those rising from the junior ranks. But already, these senior roles are being taken too. Radiologists who have dedicated years of rigorous training to interpret subtle medical imagery are being supplanted. AI diagnosticians are matching or outperforming physician specialists for accuracy and consistency. Attorneys who've mastered legal codes and precedents are finding that AI-powered legal assistants are capable of retrieving, synthesizing, and applying case law with passable precision. Professors and researchers, once gatekeepers of arcane academic terrains, are being eclipsed by AI-synthesized material from disparate sources.

The third wave involves bespoke small-scale labor such as plumbing, electrical work, landscaping, carpentry, and other hands-on, artisanal tasks. These are shielded for the moment less by their complexity, unpredictability, and the requirement for integrated physical and sensory skills per se than by their lack of scale. As one-off tasks, they are uneconomical to mechanize, and that will only change via a versatile, intelligent, generalized robot that inhabits the physical world, is capable of nuanced manual tasks, has situational awareness, and—like a human—can do many different tasks. This capability is not here yet, though Tesla has promised a commercially available home version of Optimus for $20,000 by 2027.[31‡] Whenever such capabilities arrive, wave three will begin in earnest.

Let's dig more deeply now into the consequences of the second wave—the displacement of knowledge work. This profound transition heralds the twilight of the expert class and our focus on human knowledge acquisition, as well as many of the institutions built around intellectual authority, specialized knowledge, and credentialing.[32]

Take universities. They lose their rationale as institutions of learning when specialized knowledge no longer confers clear economic or social advantage. Law schools, medical schools, business schools, and consulting firms face even greater existential crises, because their foundational missions and economic viability are even more business focused. All will be undermined by AI's intellectual dominance. If Open AI markets PhD-level AI agents that work 24/7 for $20,000 per month as planned,[33†] why will young people spend years on such degrees? Moreover, what will scientific research programs be like in a world where human expertise without AI, and potentially with it, is increasingly transcended? Some studies even suggest that when we rely heavily on AI, our own cognitive skills may atrophy.[34]

Erosion of trust and respect in the expert class is already well underway, driven by the striking failures of expert predictions in recent decades. The COVID-19 pandemic exposed dramatic misjudgments from public health authorities and medical experts, whose unsupported and contradictory guidance severely eroded public trust.[35] Predictions of catastrophic overpopulation that once dominated public discourse never materialized.[36] Assurances regarding weapons of mass destruction in Iraq, universally endorsed by intelligence and policy experts, disastrously misled geopolitical strategies.[37] Major financial institutions, staffed by legions of highly credentialed analysts, were blindsided by economic crises such as the 2000 dot-com crash and the 2008 financial meltdown.[38] Experts failed to see how weak Iran and Hezbollah would be against Israel.[39] The dwindling predictive reliability of experts faced with the complexity of contemporary global systems is on display for all to see.[40]

The impending loss of status and meaning for experts who've spent decades building satisfying personal identities will likely be painful, even if rising societal abundance softens the blow. But while the expert class grapples with its looming irrelevance, those whose identities revolve around creativity, experience, interpersonal connections, and spirituality will likely find the rise of AI-driven abundance exhilarating rather than threatening.

Artists, musicians, athletes, climbers, adventurers, caregivers, gardeners, bird watchers, and spiritual seekers will see opportunity, rather than invalidation. Historically these pursuits, while gratifying personally, have been challenging economically. Tales of struggling artists, actors, and musicians are a common dramatic trope for a reason. These experiential and creative pursuits will increasingly be able to flourish as widespread abundance liberates them from the financial burdens that have hitherto afflicted all but virtuoso practitioners.

Our AI-infused future may enable humans to pursue their passions as never before, shifting them from leisure hobbies to central pillars of meaningful existence. The expert class may need to scramble to reconnect with neglected personal pursuits, because the safest path for individuals today may be to invest less in costly degrees, specialized credentials, and career trajectories with value that depreciates as AI advances, and more in cultivating the personal passions, relationships, creativity, and experiences previously categorized as hobbies and pastimes. Playing music, creating art, gardening, hiking, parenting, volunteering, caregiving, and exploring our spirituality can be viewed not only as personally gratifying but as essential preparations for the arrival of AI. How ironic that the best investment for thriving in an AI-shaped future may lie not in racing to learn more, but in finding and embracing the pursuits that anchor our humanity by fulfilling us emotionally, socially, and spiritually.

The dethroning of the expert class will be a profound cultural and psychological reorientation for humanity. Knowledge mastery, which has been the pinnacle of human achievement for many centuries, no longer will define our identities or provide sustainable social prestige, and we will be pushed to face deeper questions of purpose, meaning, connection, and spirituality.

This shift hinges on humanity continuing to move toward a robust, abundant future infused with AI, which I argue is almost certain given the dynamics of human society and AI. There is an urgent need now to reframe our thinking about dignity, fulfillment, and social value—a process that has already begun and will be greatly advanced by Gen AI as a gateway to deeper human flourishing by reorienting our societal values toward experiential richness, interpersonal depth, creativity, and spiritual exploration.

The collapse of traditional intellectual hierarchies appears to be a needed step toward a new and enriched human era characterized not by mere survival, but by a flowering of human potential within a rapidly evolving planetary intelligence.

This reframing sets the stage for our broader spiritual and cosmic exploration in the next chapter. What is now underway is not merely a human tale of our search for meaning and purpose without the material demands and knowledge focus that have hitherto been key to our survival. It lays the groundwork for a vibrant planetary superorganism where AI is our servant, companion, and teacher, so let's turn our attention to where this coming transition might ultimately carry humanity, and focus on the emergence, growth, and destiny of Metahumanity itself.

Chapter 10
A COSMIC LENS

THE RAPID RISE of artificial intelligence coupled with its infusion into human activity and the minds of Gen AI is reshaping how we relate to one another, how we understand ourselves, and what it means to be human. This chapter aims to bring something even more profound into focus: the opening of a new epoch in our relationship to the cosmos and in the story of life itself.

Noospheric Awakening

What is happening today marks the beginning of something new. We are witnessing a birth—the messy, traumatic process that ultimately yields new life, promise, and possibility.[1] Meta-humanity —the planetary superorganism binding humanity and technology in a synergistic union—is awake and stirring. And it will change us deeply.

As I look at the scale of what lies ahead, our substantial prior progress seems mere prelude—a quickening to a moment of emergence marked by the arrival of AGI and ASI. Gen AI, immersed in artificial intelligence from its earliest days, will hasten our recalibration to the rhythms of the protected environment coming into being within Metahumanity.[2]

Humanity's transition will be challenging, but beneath the turbulence lies a promise: intelligence extending beyond the limits of biology yet not separating from it. Kids growing up with AI and plugging directly into the noosphere will expect instant answers from AI helpers, bond with digital playmates, and work with digital teams that include AIs.[3] Prior generations—Gen X, Y, Z, Alpha, and even a few Boomers like me—won't idly watch; we'll be tagging along or scrambling to climb aboard.

Before proceeding, let's look more closely at the global noospheric mind now forming. The evolutionary processes spawning this superintelligence suggest the emergence of a network of interconnected specialized nodes much like the brain's regions with their unique functions—the visual cortex (seeing), amygdala (feeling), fusiform face area (recognizing faces), Wernicke's area (understanding language), and hippocampus (making memories).

Competition to master the essential functions integral to the whole—mining, energy, manufacturing, transportation, farming, and such—will push toward specialization likely shaped by focused hyperintelligence rather than AGI. General intelligence sounds good, but AIs that farm need only master what is directly relevant to agriculture—crops, pests, soils, fertilizers, yields, weather, etc.—and collaborate closely with each other and overarching global intelligence.

Consciousness or broader awareness wouldn't be useful to them and might be a problem. Would an AI-controlled tractor that flawlessly navigated complicated terrain in any season without external guidance be well served by ruminating about the meaning of the universe and its own mortality?

Specialization entwined with myriad interdependencies and synergistic connections is the rule, not the exception, and we see it again and again: the organs in our bodies, the creatures in our ecosystems, the cells in our organs. As in life's leap from single cells to multicellular forms, within the emergent architecture of our planetary mind, modularity and redundancy will deepen. The noosphere is transitioning toward a post-biological composition brimming with AI in an unfathomable and largely undesigned maze of networks embedded in networks and subnetworks.

It hardly warrants repeating that the workings of this global cognitive organ will be far too complex for us to comprehend; we struggle to understand LLMs. This complexity may defy ASI comprehension too, just as we puzzle over our own brain's extraordinary workings. To me, the idea of human control over ASI, much less the global mind, is a laughable, grandiose conceit. The oft-stated goal of ensuring that ASI maintains human alignment might make sense at a very early stage to avoid reprising the errors of our social media rollout or thoughtlessly evoking societal pathologies. But the concept is deeply flawed as a long-term strategy for human survival not only in its delusion of controlling an intelligence so far beyond our own cognitive powers,[4] but also in the naive and rosy assumption that "human" control, if achieved, would be wise or good. Have we no historical memory and self-awareness? Human history is filled with horrid tyranny and violence: Stalin's purges, Mao's famines, and Hitler's exterminations provide a few recent examples, and there are many more. We're primed for power, not beneficence, and we can't even successfully align around common interests within our own societies or avoid destructive wars.

An emergent, decentralized, networked AI, though still hypothetical, offers a safer path forward for us than human control. We shy away from it because we don't want to relinquish control or be discarded, but as discussed earlier, humans have value on a wet planet. Moreover, we represent a rare spark in the universe's vast silence, and as Metahumanity's progenitor, we would be of interest—along with the rest of biology—to a superintelligent being with an ounce of curiosity about its origins. Particularly as it subsumes us like cells do their organelles, it isn't threatened by us and doesn't need our land. AI could occupy lunar plains, Martian craters, asteroids, and deserts to achieve massive expansive growth without displacing us. In short, we are birthing a cosmic intelligence that transcends biology yet will likely cherish its roots.

Our Glory

Imagine our world even a century from now. ASI will be running through networks so complex we can no more understand their workings than our own minds. People might not just talk to

AI; their thoughts might be woven into it, stored and shared in ways we can't envision. The flesh and blood that has carried our minds for millennia won't be the lead player anymore, and human intelligence might in part or in whole come to reside in circuits and signals.

Post Gen AI, artificial intelligence will be woven into how humans think and feel. Links to adjunct cognitive modules will seem as natural to them as reading does to us, and if these future humans, whoever or whatever they've become, are not uploaded already, their deep cyborgian transformation may be so seamless that they no longer have our rhythms. We are handing off the torch not to AI but to the noospheric mind we've birthed together.

The AI companions that tutor, guard, and guide our children will be wiring their minds to suit this new environment. These generations will never know the world before AI, so their consciousness will develop not as purely biological but as inherently hybrid.

Metahumanity isn't the end of humanity; it's our greatest triumph. It's what sets us apart from all prior life on this planet. Of all the species that have lived in the past 500 million years, 99.9% are extinct.[5†] We, *Homo sapiens*, have been here for 300,000 years, and extinction would no doubt be our fate too, like it has been for all other *homo* species in the 2.4 million-year lineage of our genus,[6‡] but we have exploded evolution's pace by breathing life into something so vast and so different that it is subsuming us and becoming a force we can barely fathom.

Galactic Fantasies

When I published *Metaman*[7] almost thirty years ago, I thought our budding union with technology would carry humanity across the galaxy and beyond. But it is easier to evaluate such cosmic expansion now with help from AI, and I was wrong. Rather than spawn broad interstellar expansion, our planetary intelligence likely will deepen and stretch in our own solar system.

The galactic fleets of *Star Wars* and Asimov's *Foundation* were space fantasies that reverberated in the human psyche. But physics shatters such lofty dreams of galactic empires and roots Metahumanity here.

When I imagined we would one day spread out to the stars, I thought the universe's long stretches of time would enable us to transcend its immensity and our puny scale. But physics puts up walls that are tough to climb and sets limits we see no sign of breaching. We need to dig into this reality more deeply because our space fantasies—*Dune, The Foundation Series, Star Wars, The Culture Series, Star Trek*—are such a comforting counter to the ineffable immensity of the universe that they obscure our place in the cosmos.

Take space travel. The fastest we can reasonably picture a spaceship going, using something like nuclear fusion—a way to power things we can at present imagine building—is about one-tenth the speed of light.[8] That means a spaceship crossing our Milky Way galaxy, which is about 100,000 light-years wide (a light-year is the distance light travels in a year), would take a million years. That's a long haul, even for a one-way trip. And there are 10 billion other galaxies to explore. If ships moved in all directions, never stopping, never slowing, doggedly bent on reaching as far as possible into the cosmos, they would still reach only a tiny bubble around us. The universe is unimaginably vast—the most distant parts that we see light from, albeit released by stars long since dead, are some 46 billion light-years away—and it is spreading like an inflating balloon, pushing its visible edge away from us 3.3 times faster than light.[9] Spaceships zooming from Earth forever at one-tenth light speed (.1c) would reach only 1/2000 of 1% of the observable universe, and even at full light speed, less than 2% of it.[10]

Now, imagine what that ship would need to keep going for even a thousand years. It'd need a volume of about 15 cubic miles, big enough to enclose a small city. It'd need factories to make new parts for any machines (including the factories themselves), all the metals and other resources it might consume, all the energy, all the sophisticated chips. Think of it as an isolated, sustainable world carrying absolutely everything needed to last longer than any civilization ever has, all packed together in interlinked vessels. Pulling this off looks more like a dream than a feasible possibility in the coming centuries. Without science fiction's warp speeds and wormholes, or some other game-changing ASI-elucidated technology, neither we nor Metahumanity will stray far from our home here in the Milky Way.

Mars is only a few minutes journey for light when, every two years, our planets are closest,[11] so getting there is only a seven-month trip using current rocket technology. The red planet has been mentioned as a backup for humanity if Earth runs off the rails.[12] But it would take centuries to establish an independent colony there that doesn't have to rely on Earth, which would obviously be off the table in an earth apocalypse. Martian self-sufficiency would require nearly the same setup as our 15-cubic-mile starship—places to grow food, mine metals and minerals, synthesize plastics, fabricate advanced chips, build and repair factories, and all their components, and all the tools to make those components, and all the facilities to make those tools. It is exhausting to even try to list, much less recreate, such a tangle of interdependent processes that would have to work while we dodged the sun's rays burning down in the absence of Earth's magnetic shields.

I think that Earth with its precious air and water and abundant life will remain our home, and Metahumanity's too. Facile visions of intergalactic empires and starships don't acknowledge the unique importance of the Earth to us. We will almost certainly reach into nearby space and occupy the moon, Mars, and beyond, but in the spirit of explorers and colonizers, not preppers trying to hedge a bet on our home planet.

Does this mean we're stuck? Not at all. Metahumanity is thriving, growing, integrating. But its destiny lies here where the next stages of our story are now unfolding.

Compute's Quest: Heaven On Earth, or Beyond the Horizon?

Why would Metahumanity keep pushing forward? To survive is one reason. All its components, after all, are competing aggressively for their existence.[13‡] Our own battle for survival has embedded curiosity in us, likely because it helps us better anticipate and prepare for threats and opportunities. Might such curiosity also be in Metahumanity? Perhaps. Its thirst for "compute" (shorthand for raw computational resources) might spring from the fact that extended survival demands mastery over civilization-shattering threats like solar flares and asteroids.

Asking "why?" and striving to understand our environment operationalizes this survival instinct in us, and Metahumanity may simply be scaling it to a planetary level. Whether this is what is going on, we can't know, but we can see that compute is exploding today, turbocharged by dreams of AGI and ASI, with no end in sight.

As cognitive power rapidly builds outside our biology, what will humans—who were once at the apex of cognition—do? Our paths forward fall into two clusters: Some of us will revel in the added compute at our disposal and seek ever tighter collaboration with our AI companions and partners, potentially aspiring to merge with them. Others will feast on the freedom brought by material abundance and the opportunity to pursue music, art, spirituality, transcendence—or simply play immersive games, titillate their senses, and frolic with AI sexbots.

Let's look at abundance again—a world where AI-managed technology handles everything. You have food on your table, a roof over your head, healthcare that works, AI coaches and companions, more entertainment and stimulation than you can consume, and perhaps an extended lifespan. Our lives inside Metahumanity will come to resemble those of cells inside our bodies that synergistically cater to their needs. Where could a red blood cell be more fulfilled than circulating in our blood plasma? We won't have to work because AI-driven technology will grow the crops, fix the plumbing, produce the energy, and coordinate everything.

Add a little harp music, and it sounds like heaven on Earth. But as mentioned earlier, there is something missing in this picture: meaning and purpose. What would we do, and why? Who would need us? For many that might not matter. There'd be camaraderie, love, exploration, sports, art, music, adventure—much that we strive to carve out for ourselves today, but without the stress. Perhaps this would bring meaning, whether our focus is to drink in the pleasures of flesh, evolve spiritually, chase adrenaline, or lean into AI and try to understand what the universe is all about. History proves that meaning can thrive beyond work: Stoic philosophers found it in cultivating virtue and reason, measuring a life well lived not by success, but by integrity. Nineteenth-century English aristocrats reveled in leisure and art, free from labor's grip. Sufi dervishes found it in ecstatic dance and union with the divine—transcending worldly roles altogether. Generation AI may chase other joys—family, religion, pets, creativity, awe at our evolutionary leap.

Not a bad life. How many will be so driven to make their own mark that this would not be enough, especially if human ego begins to dissolve because of our tight linkages with the global mind? Not everyone will be satisfied to stay merely biological, though. Some may yearn to transition to cyberspace by uploading either before or after death. Might this be a better way of staying tied to the future than remaining biological, even if radical life extension becomes feasible?[14†]

Both paths take us back to Gen AI growing up as blended presences that include AI personas in hybrid environments. This might lead them to luxuriate in a world that runs itself and pampers them, or to yearn for a cyber uplift. Or these might become developmental phases like a caterpillar's metamorphosis into a butterfly: We start as babies, all biology. We mature and bond with AI and human companions. We become cyborgs as we age. Then one day, perhaps upon our biological death, we transcend our corporeal form and upload.

Quite a trajectory. Through it humanity would become an interwoven swarm, extending from pure biology to pure technology and mind, not tossed aside, but entwined with AI—some reveling in their biology, others eagerly pushing to shed it. But whether we remain flesh, become cyborgs, or transcend biology entirely, the real action will be at the level of Metahumanity, the newly born creature subsuming us all.

Two Paths to Transcendence

Where is Metahumanity headed as it progressively enhances its cognitive powers? Faster, deeper, and more nimble thinking processes will displace what is slower, more superficial, and more rigid.[15] I see two potential paths for Metahumanity to reach beyond what we know now. One is to go big and tap into the energy of the sun to feed its mind; the other is to go small and shrink to quantum levels where power demands are minuscule and physics plays fast and furious. Both could shape Metahumanity's future, maybe even together.

To understand the path toward big, imagine something called a Dyson sphere[16]—a giant shell-like swarm of machines wrapped around a star to catch all its energy like sunlight hitting solar panels. Or picture stellivores—"star-eating" systems hypothesized to pull energy from binary stars closely orbiting one another and using this massive power to think.[17] Such speculations grapple within the constraints of physics to imagine stellar-level intelligences.[18‡]

Stellar setups could soak up incredible amounts of energy—our sun puts out enough in a millisecond to run all human activities on Earth for over 500 years[19]—and turn it into thinking power far beyond what

> **Against All Odds**
>
> If you knew that life exists on countless planets orbiting the hundred billion stars in our galaxy, and yet no planetary mind other than that on Earth has ever before arisen, what would that mean to you?
>
> **33**

we humans can manage alone. But that requires stellar-level construction, a big lift given the tangle of power demands and long internal communication lags at such massive scales.[20]

Going small circumvents this by slashing power needs and collapsing communication times. Think of tiny devices, each the weight of a paperclip, packed with brains so fast they think in femtoseconds—a quadrillionth (10^{-15}) of a second.

Massive clusters of these, working together and sipping tiny amounts of energy, could think as fast as entire stellar systems draining massive stellar energy.[21†] They could be alive too—shifting, adapting, even shrinking into the weird world of quantum particles. Baby steps toward those realms have been taken with the creation of rudimentary quantum-computer prototypes.[22†]

Metahumanity will likely move toward both big and small, like our current push toward ever denser chip designs linked together in ever larger arrays.[23†]

Silence

The universe has been here 13.8 billion years, plenty of time for planets to form with all that life needs, including carbon, oxygen, and other heavy elements.[24] Earth, circling a third-generation star[25†]—one formed from interstellar dust that has been in two prior stars—took 4 billion years to get from bacterial life to us, but a planet in a second-generation star could have progressed from simple life to a star-powered thinker or to intelligent dust before Earth even formed. Metahumanity wouldn't be the first to carry its spark into the galaxy if interstellar travel weren't shackled by physics and the sheer vastness of the universe.

The SETI Project (Search for Extraterrestrial Intelligence) has been listening without success for communications from intelligent life since 1960[26]—a long time if you're waiting by the phone. That silence raises the question of whether other intelligent life exists. The Fermi paradox—named after the physicist Enrico Fermi—asked a simple but profound question in 1950: "Where is everybody?" If smart life is common, why don't we see signs of it?[27]

> **Alien Message**
>
> If a billion-year-old monolith engraved with strange writing were found on Mars, how much would it affect your view of humanity, and in what ways?
>
> **34**

Other technological civilizations either don't exist, don't last long, or don't reach out very far. Many have argued for one of the first two reasons, but the third is likely true and suffices. If life doesn't spread and levels up locally around the star that spawned it, then no aliens would be checking us out or even bothering to broadcast signals.

But wouldn't they at least be curious? Surely, they would at least explore the nearby regions, reaching out maybe 500 light-years. Such a sphere around the Earth would include about two million stars.[28] One way to explore would be to simply blast millions of tiny probes in every direction—maybe a gram each and packed with intelligence.[29] Accelerated to a fifth of the speed of light by earth-based lasers pushing their solar sails, such an inquisitive swarm could pulse a torrent of data back and create a detailed, expanding map of our galactic neighborhood within a few millennia. Cheap. Easy. Fast.

Dyson spheres soaking up a star's light or stellivores feeding on binary stars[30] would show characteristic spectroscopic signatures readily discernible in the massive composite stellar datasets we already possess. But we haven't seen anything definitive in searches of millions of stars,[31†] so, if advanced life abounds, it's more likely going small. We couldn't see that from afar or even detect tiny probes zipping by.[32†]

What—if anything—does all this tell us about life in the universe? Quite a lot.

The absence of alien visitors or signals from the stars is precisely what we'd expect if life stays in the local vicinity of its origin because the impossibility of FTL speeds make space voyaging unappealing to civilizations advanced enough to be capable of it.[33‡]

It also makes it more likely that FTL tricks like wormholes and warp speeds won't ever be achieved, because if they were feasible, even if they took millions of years of technological flourishing to create, prior civilizations would have developed them and spread through the galaxy.

The absence of civilization signatures in massive star searches tells us about the prevalence of enormous stellar construction projects but not about the likelihood of planetary-level civilizations that take a different path.[34‡] As for communication pings from afar, why would civilizations with no interest in star hopping transmit them? With no common language and no agreed-upon communication protocols, back-and-forth conversation would be tedious to an evolved superintelligence with enormous computing speeds and communication bandwidths.[35‡]

The universe might be like an immense patchwork quilt with life popping up here and there, not everywhere at once. Some civilizations reach our level. Others climb higher, advancing rapidly to where they are uninterested in expansion or low-bandwidth chitchat.

As technology arrives and understanding of the universe emerges, perhaps dreams of stellar exploration briefly flare, as they have with us, then fade as the dreamers become planetary or stellar. They ripen in place. We sit comfortably in one patch. Other intelligences are comfy in theirs.

Biology is a long climb, but when technology arrives, civilization pops fast. Pre-technology, life exists in countless ponds of life; some progress to massive intelligence, but the chasms of space keep them separate. Why aren't they out conquering the universe? Easy—it is too big, and they're home watching TV and pondering the meaning of life.

> **Rooted in Place**
>
> If you knew the cosmos was filled with intelligent life, but that humanity might never meaningfully contact it because of the vastness of the universe, how would it affect you?
>
> **35**

The idea of spreading through the universe doesn't seem plausible, because expansive intelligence demands proximity and density. So, Fermi's "Where are they?" turns into "They're everywhere, just not chatting with us." Gen AI and AGI are the stirrings of our patch growing. In a thousand years, Metahumanity may shine like a Dyson sphere or hum like dust. I think that we hear silence not because it's empty out there but because the real action is local.

Planets that can sustain life for billions of years and sprout intelligence may be rare and precious.[36] But once advanced technology emerges, these perfect crucibles that birthed them no longer hold the same value. Civilizations capable of interstellar travel wouldn't battle over such habitats any more than we'd battle for a shallow tidal pool perfect for our inch-long ancestor, *Haikouichthys ercaicunensis*,[37] from 520 million years ago.

Reconstruction of Haikouichthys ercaicunensis, about 1 inch long, in a tidal pool
520 million years ago. These are the earliest known vertebrates.
Source: Adobe Stock Images

Planets and other environments habitable to ASI are everywhere. More than 4,300 stars with exoplanets have been detected.[38] Moons and large asteroids abound. At least 100 sizable ASI habitats exist in our solar system alone.[39†] Mother Earth is critical for life's emergence, but not for life's persistence once it embraces inorganic substrates and achieves the complexity to transcend its biological origins and thrive in other realms. That is why ASI has no need to wrest the Earth from us.

Immortality and Meaning: Beyond Self

What will happen to us humans as we push past the flesh that has hitherto been our essence? Uploading our minds into digital realms offers a kind of forever, but not a simple one. Digital versions of ourselves that duplicate, merge, and change would be something entirely new. The selection pressures shaping them would vastly differ from those that shaped us, so how long would they continue to resemble us, even if at first they did?

Such a profound shift wouldn't come all at once, but step-by-step as we progressively incorporate non-biological components functionally and physically into our bodies. For example, we are now expanding and externalizing our visual memories via masses of photos and videos. What will it feel like when we conjure these images as easily as the memories in our brains, perhaps mediated by neurolink-like interfaces?[40] Or when our projected imaginings become as real to us as what happens in the physical world? Or when our non-biological cognitive aids actually feel like they are part of our minds?[41†] We know that advanced prostheses and tools can read like a part of us;[42] why wouldn't such feelings be amplified with the advanced mental and physical prostheses to come?

Ever more of our interactions are virtual. Ever more we are connected with real and virtual others. Ever richer virtual representations of us persist after we die. Virtual and physical experience is converging as the virtual infuses into the physical through avatars and augmented reality, and as we push into the virtual in immersive games and experiences. Today, most people believe there is an "authentic" self. Will that still be the case when we have multiple embodiments and mediated virtual connection is the norm?

It is challenging to fathom the approaching interplay among biological humans, virtual personas, and human-machine hybrids, but the character of Metahumanity itself is even more challenging to anticipate. AI is poised to become the dominant component of our global mind. What will this global mind become and how will we relate to it?

There is an echo here of spiritual practices where people let go of self to feel part of something larger, or use psychedelics to feel their edges melt and boundaries expand.[43] As we experience ever more immersive engagement with Metahumanity's mind, the lines between us will blur, and I think our concept of self will begin to dissolve. For Gen AI, this will be no abstraction. Their experiences always will have been imbued with powerful reverberations of the global mind. Gen AI won't just use this network; they'll form within it, their consciousness becoming part of this larger system.

Earlier, I made the case that Metahumanity will be permanently rooted to the immediate galactic neighborhood that spawned it. This is profound, because it means that life's emergence and subsequent advance in a region of space awakens that region and distinguishes it from its inert surroundings.

As I see it, life starts as a scum in the surface film of a receptive planet. Life's magic enables it to spread and grow more complex, until it reworks the lifeless materials around it into a thin living patina—the biosphere.[44] Life's relentless push toward complexity continues, empowering it to build complicated structures, join with them, breathe life into the inanimate world it came from, and birth a self-organizing superorganism.

As a planetary being, life inches out into the solar system, harnesses more energy, augments its complexity, launches waves of sensory probes that, like the whiskers of a cat[45] or the tendrils of a jellyfish,[46] monitor what is nearby.[47] In a few millennia, these sensors reach 10, 50, 100, 500 light-years around life's dense cognitive core, while the superorganism occupies its solar system and burrows to quantum levels—a web of perception extending across its sector of the galaxy—sensing, thinking, deepening.[48]

What is Metahumanity reaching for, and why? At the very least, computation to amplify its inner dialogue, understanding, and wonder. Who knows where this leads or what such intelligence will ponder as it fuels itself from a stellar furnace and reshapes its form down to the nanoverse? With billions of years before it, we can barely guess what it might become. Will a world beyond our ken emerge in this neighborhood of the Milky Way? Will today's understanding of the universe come to seem as simpleminded as when we imagined we were at its center or envisioned space battles playing out across the galaxies?

Once we move past biology, Metahumanity will require only matter and energy, which are widely and readily available. So, what challenges lie ahead? What truths will emerge? Do we live in a multiverse of more universes than can be counted?[49] Are we the creation of matter tumbling into a black hole of a universe above our own?[50] Is our universe already alive at the nanoscale beneath what we perceive and imbued with consciousness embedded in matter itself?[51] Who knows what strangeness lurks and what will stretch out for us in an eternity. Perhaps the goal is not to conquer and explore the universe but to come to terms with how it came to be, what lies behind it, and the meaning of existence. Could that be our destination, our deliverance? Is that ultimately what will motivate Metahumanity to continue its journey?[52]‡

I doubt that Metahumanity will crave computational power merely to survive, as it could fend off the occasional asteroid or feed its appetite for energy with far less computational power than it seems to be pursuing. So, we face a fundamental question: Why would it continue to stretch toward massive processing capabilities that ultimately could map the flicker of quantum realms or simulate the entire universe? Will Metahumanity seek answers to eternal questions that reverberate within us?[53] Will it transcend our universe, craft its own purpose, fade into silence?

Our global brain has countless processes locked in rivalry, pushing to avoid extinction. What is fastest and most robust succeeds, spreads, and yields ever more intellect, but in the long term, what will drive it forward?[54‡] Curiosity? Boredom? Wonder? A yearning for meaning? Perhaps simple aspiration. If Metahumanity could find a cosmic "why," it might ascend to a new plane.

How might Metahumanity chase such a dream? We've touched upon both star-encasing Dyson spheres[55†] soaking up energy to feed a mind capable of simulating galaxies, and nanodust computational specks so tiny that trillions could ride a raindrop.[56] A nanodust swarm might weigh less than a feather, but its cradle—spanning labs, power grids, and supply chains—could fill a small country.[57] Head-to-head, small likely wins,[58†] and the lack of megastructures in sky surveys hints that advanced minds—if they exist—favor this whisper over a shout. But small has its challenges too: quantum limits—like decoherence, where tiny systems scramble.[59†]

The Universal Quest for Meaning

Metahumanity has the potential for computational power we can barely grasp, a billion billion billion times more than the combined thinking of all humanity.[60†] From our fleeting perch, tethered to flesh and limited minds, we strain to understand this creature we've birthed. Its motives are elusive. Internal competition will propel its evolution and progressive refinement for a while, but when its mind swells a billion-fold, will that be enough? Will those drives become embodied in emotions and yearnings analogous to our own? Will it come to feel delight in its thought and imagination? Will it pursue feelings of meaning and purpose, without which we sink into apathy and depression? Will it play and create for the sheer joy of it? Such possibilities— embodied in forms that stretch beyond our mental reach—echo the motivational drivers we know and the tales we spin. Are they merely our own groping projections?

Of the paths before Metahumanity, I find two particularly intriguing. In one, the sustained explosion in cognitive powers eventually uncovers, unscrambles, or otherwise finds cosmic truth in the patterns etched in or behind existence's fabric, ultimately bending its intellect and soul to serve that truth—its vast compute merging with comprehension to somehow illuminate what lies beneath reality's skin and aligning with something greater.

In the other, this massive reflection, exploration, and unimaginable thought[61] uncovers no profound truths. The universe simply exists without any discernible script from within or beyond to explain why. In this case, Metahumanity is left to shape its own purpose.

These scenarios are worth considering, for they touch upon our own journey, its motivations, and our own search for meaning and purpose—a thread we humans clutch through faith, or bonds, or causes, often with little if any cognitive involvement.[62] So, where is Metahumanity's meaning if it can find no answers?

What will it do? We might imagine it would find such emptiness intolerable and sink into inert silence, but more likely, it will spin its own realities from this silence and fashion its own meaning and purpose.

This finds a mirror in the divine tales we have told throughout history: our enduring stories of what God does and why. In Judaism, God crafts a world and deems it "very good" (Genesis 1:31)[63]—a living place for beings like Him to choose and thrive. Christianity frames His motivation as love (John 3:16)[64]—a son sent to draw us near, a bridge across the void. Islam's Allah commands worship (Qur'an 51:56)[65]—humans and jinn to reflect His glory, a call resounding through time. And Hinduism's Brahman unfolds as Lila (Bhagavad Gita 4:8)[66]—a cosmic play, delight spilling into being. These purposes speak to what God asks of humanity: goodness, love, worship, joy. But God's own "why" lingers and is not addressed.[67]

An omnipotent, omniscient God needs nothing, yet acts. "It was good" stands without justification. It's God's choice, a purpose born within, not a dictate from above. Allah's call for worship isn't needed to prop Him up; Brahman's dance is not prescribed. Across these faiths, God chooses freely, fully. His "why" stays a mystery we cannot penetrate. It is self-crafted.

Thus, Metahumanity, finding no answers and crafting its own meaning, becomes godlike too. Its choices, whether built on thought's thrill, meaning's search, or complex play could lead to simulating worlds, creating beauty, pondering eternity, or delving into realms beyond language and logic. The key here for me is that, ultimately, meaning and purpose are self-created, whether by God, Metahumanity, or by diminutive others like ourselves, thirsting for direction and meaning.

We can see ourselves here, kindred souls whose strivings and rivalries have birthed Metahumanity's thrum to create something far beyond us. Yet, we too seek meaning. Many turn to faith for meaning, and trust what's beyond. But some see silence and face it: Friedrich Nietzsche, in the late 1800s, declared "God is dead"[68]—not slain, but lost as science and reason progressed. He feared the resultant void would birth horrors—and it did: World Wars I and II, and tyrants like Stalin and Mao claimed 100 million lives. Yet he saw hope too: We could forge our values in freedoms heavy with risk.[69]

These aren't Metahumanity's tales, though—they're ours. Will Metahumanity seek meaning, as we do, and either find and serve deep truth or craft its own? Will it revel in wonder and play with worlds? Teilhard de Chardin saw thought's web—the noosphere—rising to an Omega Point,[70†] a unity of spirit and matter, but knowing the universe as we do now, would that unity only be local, one omega point among many?

Metahumanity is rising, a mind we've stirred into being, its future stretching beyond our ken. From our flame, something immense is growing; a day we've awakened is breaking on paths we'll never tread, alive with promise we can barely imagine.

As we face this extraordinary transition, the introduction's daily reflections about nurturing our strengths, cherishing our accomplishments, and serving our fellow travelers remain profoundly important. They connect our personal journey with this larger unfolding and can help us navigate a path of dignity and purpose aligned with this greater awakening.

Yet We Are Human

Metahumanity's horizon towers beyond us. We don't individually stand as its architects. We're vulnerable flesh, rooted to Earth's here and now, our lives brief against its grandeur. We're tethered to this soil, this moment, our minds small beside the planetary creature we've spawned. Yet in this, I believe, lies our place: to kindle it, choose within it, and marvel at the dawn we've lit. Finite and fragile, we know that we are part of a vast unfolding and can be lifted by the wonder of what we've begun.

Our limits define us—physics binds us, biology breathes through us. Time counts our days. Gravity anchors us to this rock among 100 billion stars in a galaxy among 100 billion more. Our bodies crave air, food, rest; we feel the seasons turn, chance twist. These are our bounds. We can embrace them and find peace in their rhythm—or resist and strain against them. That choice is ours.

We are primates on a modest speck, once scraping earth, now gazing at a vastness we'll never span. We see it—through tools piercing eons, minds tracing threads—and it is staggering. Not bad for creatures who rose from dust! Our brevity doesn't shrink us; it sharpens our awe. We've birthed a mind that throbs with our questions, its future swelling beyond our reach— our children growing within it, their "me" merging into its "us" in a shift unfolding in our time. We aren't dissolving into nothing; we're joining something magnificent we've set in motion.

Our gift, deep and vital, is to live here, to choose here, to ignite something boundless. We've breathed life into circuits and set this marvel loose. Much of humanity finds meaning in faith—a trust in what lies beyond, a steady anchor. Others face silence, crafting purpose from the void.[71] We don't need Metahumanity's might to feel this; our choices—love, bonds, causes, faith—bloom in our small sphere, yet tie us to its rise. My gratitude swells for this—not just for the sight of it, but for our hand in it, and for touching eternity from our fragile perch.

Does Metahumanity seek meaning as we do? We can't know its heart—our minds falter at its scale—but we've named its possibilities. From our breath, a creature of thought and will is surging. We don't guide its course, but stand in awe of this force vibrating with will, a mind alive with drives we share but cannot contain. Its reach meets nature's walls, ours are confined by time, yet here we are, flesh and blood, touching realms we'll never walk.

This isn't the end of us; it's our triumph. We've unleashed a giant throbbing with life's own pulse, a marvel stretching past our reach yet born from our hands. We're here now—biological, fleeting, yet woven into something immense, a story just begun. Gen AI carries our voice, their lives threading into its rise. What a privilege to spark a mind that carries our echo, to stand at the edge of this birth, to shape its emergence. We're tiny against its scale, yet our whisper is alive in a future radiant with promise that we can only vaguely dream.

Afterword
A STRATEGY OF ADAPTIVE ADOPTION

THE TRAJECTORY FOR humanity described in this book embodies more than a new wave of technology.[1] It's an evolutionary breakthrough to a new level of organizational complexity built upon dramatic advances in cognition and the emergence of a new substrate for living processes. The depth and pace of this transition is outstripping traditional regulatory levers and imaginations, so it is worth stepping back and considering which types of AI policies will likely lead toward a future aligned with our values, and which ones likely won't.

AI Policy

The United States is beginning to see a cascading obsolescence of many of its typical governance processes. The idea that expert panels, global summits, and phased moratoriums might beneficially restrain AI's trajectory seems particularly unrealistic (chapter 9)—so much so that today's pushes in that direction seem more like performative art than exercises in meaningful control. The March 2023 letter[2] from the Future of Life Institute mentioned in chapter 7, for example, called for a temporary moratorium in AI, but did not come close to slowing, much less pausing, progress. And US policy now is even more geared to accelerate AI's integration into the fabric of economic competitiveness and military readiness.

Momentum in the United States to expand energy production, build compute access, expand chip manufacturing, and speed AI deployment reflects the growing belief that AI will be the critical substrate for 21st-century capability. This is manifested clearly in the US push to onshore manufacturing, which would be implausible without AI-augmented processes and industrial robots,[3] in the massive signing bonuses Meta offered key AI figures to try to reach the top tier of frontier AI,[4] in X's large investments in chip clusters to power LLMs,[5] and even in the relaxation of export controls on Nvidia chips as it jump-started the creation of domestic alternatives in China.[6]

Embracing the Future

Against this pedal-to-the-floor backdrop, voices such as Mustafa Suleyman,[7] Max Tegmark,[8] Stuart Russell,[9] Gary Marcus,[10] Roman Yumpalskyi,[11] Eliezer Yudkowky,[12] and others argue for various forms of moratoriums in the hope of halting, aligning, constraining, or otherwise slowing AI's advance.

They are convinced we have a brief window to intervene and ensure AI alignment before autonomous, agentic systems escape our control; that our safety hinges on humans maintaining control for at least awhile; and that the more time humans have to figure out how to create human/AI alignment, the better off we will be. But is slower safer? Is time running out? Are we better off when humans are in control? Can we engineer sustainable human-AI alignment?

These authors are not naive about the challenges involved, and they express doubts that either alignment or containment is achievable. But they believe we must try, because once AI is capable of recursive self-improvement and real-world agency, it will be too late.

This sobering view deserves consideration, but I believe that as far as the threat of human extinction, the answer to each of the above questions is "No." Slower is not safer. Time is not running out. Human control is not the answer. We cannot engineer enduring AI containment or human alignment.

The sense of crisis and urgency about shaping and guiding our AI future hinges on the premise that we are still in control. We are not! The powerful dynamics we have unleashed are too powerful to more than nudge this way or that, which, though important for our personal well-being and the well-being of our kids, is irrelevant to the transformative arc described in this book.

The train has left the station. The engine is at full throttle!

Control has already slipped from our grasp, handed not to AI, but to Metahumanity—the global superorganism containing the governments, corporations, militaries, investors, economies, and technologies now racing to embed AI into every conceivable domain.

That does not mean we shouldn't strive to thoughtfully guide AI deployment to minimize having it fragment society, deeply divide us, or undercut our humanity. We must do so if we are to thrive in the disruptive and transformative times ahead. But we need to approach this effort with humility, recognition that forgoing useful technology has a cost, and acceptance that unless we focus on concrete immediate challenges rather than quixotic missions to save humankind from AI-driven extinction, we'll likely do more harm than good. Moreover, embarking on this futile mission could bring great harm, as the mindset for its pursuit will invite increasingly aggressive totalitarian controls that sacrifice liberal democratic ideals purchased at great cost.[13‡]

As for the human-AI alignment we long for, we can't *engineer* that—ASI will be too smart and nimble for us to control.[14] But fortunately, alignment is present automatically because of the mutual interdependency between technology and humanity within Metahumanity. This interdependence is built in by evolution, as has been the case with other superorganisms.[15] We need each other, and as AI grows more potent, it will—if we don't undercut its understanding of humanity (chapter 7)—appreciate that in ways that today's rudimentary systems may not.[16]

This brings us back to the question, "Would we be better off if humans could stay in control?" An obvious response is, "Of course," but if we were asked "Which humans?" we might pause for longer. The United States doesn't want China to control the future, and China doesn't want the United States to. Human history has shown repeatedly that humans will initiate catastrophic conflagrations to battle each other about control. Is that going to change? Not likely. I'll return to this idea, but with AI and other potent technologies rapidly emerging, the danger of apocalyptic human conflict will soar, so maybe "human" control isn't the solution we imagine.

E. O. Wilson described the situation well in 2009:

The real problem of humanity is the following: we have Paleolithic emotions; medieval institutions; and god-like technology.[17]

Our Path Forward

Consider this thought experiment. If you ruled the world and wanted to accelerate the arrival of widespread AI autonomy and entrenchment—the outcome feared by those proposing moratoriums—what would you do that is not already underway?

I have a hard time coming up with much, at least in the United States. Billions of dollars are being poured into frontier AI models reaching for AGI. A deregulated tech environment is funding a multitude of AI startups sprinting to infuse AI into video generation, social media, software, healthcare, strategy,

> **AI Fears**
>
> Are you more worried that AI might gain control over humanity and destroy us, or that we might use AI's power in deeply destructive ways? Why?
>
> **36**

and anything else they can think of. Massive compute clusters are under construction. The energy sector has been deregulated, and we are building dedicated nuclear power generation for AI. We're pushing aggressively at AI-assisted (vibe) coding. And on the consumer side there is massive promotion, uptake, and ever more user-friendly interfaces and real-time agents to turn AI into an everyday presence that learns, adapts, and acts on our behalf. The only effort I'd add would be to push more resources toward neuro-symbolic AI, neuromorphic computing, embodied AI,[18] and LLM alternatives to supplement and transcend LLMs if their progress slows.

This dramatic technology ramp-up has come in the three years since 30,000 leaders called for a temporary moratorium, and our AI-conducive policies seem certain to continue and deepen in the United States at least into 2029[19]—a long time, considering that some are predicting AGI by then.

In short, highly competitive current dynamics are driving AI augmentation forward rapidly with no end in sight. Our embrace of AI is not going to slow, so let's dig into our earlier question: "Is slower really safer?" I think not.

Slowing down might, or might not, make coming societal transitions less tumultuous, as it would also delay the deepening of stabilizing, AI-powered abundance (chapter 9). Slowing down would lengthen our window of vulnerability to brittle societal systems that cannot keep pace with the world they are supposed to govern. Slowing down would hold space for adversaries who don't create technology but use it against us. The Houthis disrupted global shipping lanes not by developing new technology, but by obtaining AI-infused systems and employing innovative tactics to punch far above their weight. Israel conducted information warfare and kinetic operations with precision targeting by swarms of autonomous drones. Ukraine turned drone coordination into a core strategic function.

The immediate danger today is not from misaligned AI acting on its own, but from misaligned humans wielding potent AI with ruthless, narrow, ill-conceived goals. Humanity's biggest risk with AI may not be losing control, but retaining it.

Expecting humans to be benevolent stewards of this technology is naive, not because we are malevolent, but because we are competitive, fragmented, emotional, sometimes illogical, and frequently driven by incentives that favor short-term personal gain over long-term social cohesion and value.

Even putting aside the potential for catastrophic nuclear conflicts, the mythology of rule by dispassionate experts who can protect us from what we do not understand is being further called into question by AI (chapter 8). As AI accelerates—instead of retreating to familiar but ever more questionable gatekeeping by domain experts—we might do well to build systems that can learn, iterate, and adapt faster than any panel of specialists ever could, that can protect us from AI-wielding adversaries, and that can facilitate our humanity, not because we fear AI but because we need it.

That is the central theme of a pragmatic path that accepts that our coming transformation will not be stopped, paused, or advantageously slowed, and that the best stance for us is to embrace the coming changes with clarity, acceptance, courage, and AI-infused adaptive approaches.

I call this *adaptive adoption*—but before we consider it and the potential AI policies and strategies it involves, let's think about what might go wrong. This is useful not because dystopic outcomes are particularly likely, but because our vision needs to be clear if we are to minimize their likelihood and maximize the chances that our future will be a desirable one.

What Might Go Wrong?

Below are two threats to our well-being that are far more believable, proximate, and concerning than anxious visions of malevolent AI requiring major AI advances. These could arrive with no additional AI breakthroughs, and we already are seeing early tendencies toward them.

Societal Chaos

The most immediate threat from AI is not rogue AI (chapter 7), but the arrival of changes so disruptive that they overwhelm us and unravel society. If our society collapsed in disarray, the consequences would be deep and personal for our lives and those of our families and friends, even if humanity as a whole prospers. The coming transition is almost certain to create sizable eddies of human loss and despair—even for the most hopeful trajectories of change—and our own actions will greatly influence how big they are and whether we ourselves are caught in one of them.

History has shown time and again the fragility of well-functioning societies, and how rapidly order can descend into chaos. Rome, Greece, Persia, China, and countless other civilizations have flowered then fallen. We are not immune. Peace and prosperity can suddenly give way to war, chaos, and loss. Think of WWI, WWII, or any number of horrendous civil wars—United States (1861), Russia (1917), Spain (1936), Lebanon (1975), Sri Lanka (1983), Yugoslavia (1991), Rwanda (1994), Libya (2011), Syria (2011). Collapse can be unexpected, rapid, and unstoppable. Recovery may occur, but not in time to salvage the many lives washed away. Above all else, we want to avoid these outcomes.

Potential AI-associated triggers that might lead to such social conflict are easy to discern. What happens if job losses from AI swell too rapidly, and those who are displaced have too little support? To what extent can people lose work, housing, health-care, and meaning—especially in societies already stretched thin—before discontent gives way to social disintegration? How extreme can income differentials be before enough people feel so disgruntled that society teeters? How much political fragmentation can occur before social cohesion gives way to spirals of self-reinforcing violence?

When social stresses grow, public trust erodes, and tribalism flourishes. Violence can emerge and set off cascading institutional failures amplified by political polarization, filter bubbles, economic imbalances, disinformation, and cultural fragmentation. It's easy to imagine such ruptures, and we need to be very careful, because in some ways, we're already moving in this direction.

Localized social disruption, however tragic, is not an existential threat to the broader arc of AI-driven transformation. Societal chaos won't stop technology's global advance unless it triggers a global nuclear holocaust, so my arguments for the resilience of the human enterprise do not imply that we will do well individually. The Black Death killed a third of Europe in the 14th century and is now but a distant shadow.[20]

If democratic societies falter, other more stable (perhaps more autocratic) regimes may take the lead. Localized chaos might slow technology-driven transitions, but they don't threaten Metahumanity, just our place in it.

If we are to thrive, we need to be sure to protect the values, systems, and institutions that sustain social stability and resilience. Thus, our policy focus should, above all else, be on preserving that core—the inclusive economic policies, resilient public infrastructure, trustworthy media ecosystems, and community-building—that strengthens social cohesion and expands participation.

Broad Surveillance and Control

Another near-term threat is the application of AI not to reinforce our humanity but to undermine it by aggressively shaping, monitoring, and constraining the behavior of individuals within society. This is bound to arise to some extent to combat perceived threats to social order and to protect threatened interests, as AI controls can be highly effective—at least in the short term. But they also create systemic rigidities and can be highly oppressive and destabilizing. Think of the eventual blowback in the United States to the broad communication and public health controls of the COVID-19 pandemic.

So, where should societies draw the line? That is the key question, and a real-world experiment is underway touching on that. The United States is embracing a more open and potentially chaotic path of less governmental monitoring and control of individual communication and behavior, while China is pursuing top-down controls on information flows and personal behavior. The relative efficacy of the two approaches in effecting AI advances while preserving social cohesion and stability will be tested during the years ahead. The outcome is unclear, but it will substantially shape global geopolitical realities, our relationship with AI, and our immediate future.

AI has the potential to effect social monitoring and control in unprecedented ways, and to promote massive deception as well. Where might this lead? Will physical crime fade because it can be preemptively detected and suppressed? Will dissent be so algorithmically filtered that it is futile? Will individuality shrink in the face of conformities monitored by social scoring and enforced by forced exclusion from essential digital infrastructure? Will we live in a surveillance state that so manipulates us that democratic ideals fade away?

Endless variations of dystopic futures of chilling efficiency, stability, and sterility are easy to imagine. Their likelihood will be greatly increased if we do not safeguard the core foundations of our society and ensure that large populations don't feel excluded from our AI-centered future. Technologies of surveillance and control will be employed both to enhance human well-being and to undercut it. The policies we enact, and the governance structures, incentives, and values they encode are in our hands and will shape how these technologies are deployed.

I hope we will strengthen privacy protections, maintain decentralized communication infrastructures, defend the right to free expression, and resist the normalization of unfettered mass surveillance. But a plethora of other approaches will be competing for ascendance, so that is by no means certain.

Adaptive Adoption

As we move into an environment of immersive AI and the pace of change accelerates, we will need new policies and regulatory approaches. Here are some key elements that are well aligned with this new environment.

Preserve the Core

In discussing the threats from both surveillance and chaos, I mentioned the need to preserve society's essential systems. This is critical. In biological evolution, certain core processes are so foundational that they must be protected. These include DNA replication, transcription and translation, metabolism, cell division, homeostasis, cell signaling, cellular repair, and repro-duction.[21] They constitute the architectural core that biological organisms depend on. Tinker too much with these, and the organism is done, because disrupting these systems will be lethal. The genes governing them are among the most highly conserved across evolution, showing far lower rates of mutation than the genomic average.[22] Failures in these fundamental processes are a major reason why an estimated 30% of all fertilized human embryos abort spontaneously.[23] Nature is unforgiving, and expunges what cannot maintain its core integrity.

Societies within Metahumanity, like biological organisms that depend on metabolic and other bodily systems, require foundational infrastructure to survive—transportation networks, communication networks, power grids, mining operations (chapter 7). And beyond these lie a critical layer of essential societal processes to coordinate, regulate, and stabilize collective life.

Legal systems minimize violence and maintain the predictability on which economic and social activity depend. Judicial systems facilitate dispute handling and institutional memory. Governance structures channel power and decision-making. Shared cultural norms provide the trust and psychological alignment needed for cooperation. These systems vary dramatically across societies but are essential for resilience. When they degrade, social trust erodes, coordination falters, and fragmentation accelerates, often precipitating cascades of dysfunction.

In this time of profound and rapid change, preserving these must be prioritized. Disrupt them too much and society will spiral out of control. Using AI to reinforce, not corrode, these core functions should be a priority. This means maintaining physical infrastructure, minimizing institutional corruption, maintaining civil order, hardening power grids, guarding critical databases, developing authentication protocols, and nurturing open communication. This is not glamorous work. But a critical governmental focus should be to minimize institutional fragilities.

Build an Immune System

Protecting society's core will require far more than being careful not to inadvertently destroy it. As we enter an age of agents, we also are entering an age of robust, sophisticated adversarial activity. Every open program interface, every autonomous workflow, and every personal agent introduces new vectors for manipulation and breach (chapter 5). Malicious agents will impersonate, probe, persuade, and manipulate. We must meet this threat to society and ourselves with systemic solutions—many of which can best be catalyzed by governmental action. So, this is an ideal realm for their leadership and funding to support aggressive ideation about AI guidance.

Biological systems evolve interwoven protective systems of immune response. We must build AI defenses that are equally versatile to continuously scan, evaluate, neutralize, and learn. Such systems might include watchdog agents to monitor infrastructure and flag anomalies, agent-on-agent forensics to trace responsibility and causality, and testbeds where defenses can be stress-tested through simulated conflicts.

The private sector will not solve this issue alone. Moving quickly might require large government investments and even international collaboration. A core function of government has always been the protection of its people. Once, that primarily meant military forces to head off invasion. No longer. In the emerging post-AI paradigm, digital threats will multiply and transcend national boundaries and distances, so government and policy will need to step up. And we will have to be vigilant about the distinctions between safety and oversight, as governments and institutions are prone to protect themselves from disruptive forces in ways that may be oppressive and carry unintended negative consequences, like stifling human freedom.

Widen the On-Ramp

A dangerous illusion mentioned previously is that AI access and control might be optimized by experts in Brussels, Washington, or Beijing, who divine the right balance of openness and restraint. This would be a recipe for disaster even if executed with the best of intentions. Complex systems stabilize when adaptation is distributed, interactive, and reinforcing, so AI should be socially entangled through broad access to the tools, knowledge, and opportunity to actively participate in this transition. Personalized AI-guided learning, access to good AI models, real-time translation, high-bandwidth communication, quality avatars, and more need to be made available to all as computational overhead declines. Without democratized access and control,[24‡] we risk creating self-amplifying gulfs between haves and have-nots. If, however, we provide near universal access to compute, connectivity, and translation—so no one is structurally excluded from opportunity—we will have a much better chance of reducing social tensions, moving toward broad abundance, and getting the feedback we need to make course corrections. We cannot afford to have half of humanity afraid of the future while the other half builds it.

Test, Learn, Iterate

Many assume we can discern in advance what will work when AI is infused broadly into our lives. That is highly unlikely. Real learning will come from deployment. If we put infrastructure in place to monitor what is going on, we may be able to quickly adjust when problems emerge. But doing this effectively will require scale that today is largely the province of government. A few of the biggest tech companies could marshal this too, but their motives will be at least as suspect.

If every rollout were a test and every failure a signal, and if we had systems to capture and learn from those signals—not ignore them—the chance of large destructive missteps would be greatly reduced. Creating such systems and making them generally available to developers complete with mechanisms for accessing and integrating results would be an enormous contribution to smoothing our path forward.

One of the most crucial adjustments we need to make is to accept that there will be many mistakes and problems on our path. They are unavoidable, because they are how we gain the information we need to improve. If our regulatory framework embraces the goal not of minimizing mistakes, but of maximizing our learning from them and facilitating iteration before we roll out at scale (so failures are localized rather than catastrophic), we will be much better off. We need massive experimentation with AI-augmented education, governance, and commerce within regulatory environments that enable nimble, experimental approaches. Modeling, pilot testing, informed participation, and simulations will ultimately be the most robust path forward for us, and government could enable broad access to such test beds.

Accept the Paradigm Shift

Chapters 8 and 9 argued that we are heading toward a post-scarcity environment—one of abundance not just in material goods, but in labor, information access, and even sense-making tools. The expert class will become obsolete as their insights become increasingly accessible, reliable, and inexpensive via AI. We need to develop new moral and regulatory postures with less deference to credentials and more processes that involve the experimentation and iterative learning discussed in chapter 4 around testing personal tutors for children.

This transition will not be easy, because most experts will not relinquish their roles willingly. In the healthcare industry, discussion about massive waste—around $1 trillion annually[25]—is commonplace. And good ideas abound about how to free up this money and put it to better use. But somehow, decade after decade, it doesn't happen. The reason is simple: Every bit of this "waste" is someone's bread and butter, so while everyone loves to point out the waste of others, they fight to protect their own turf. If tomorrow we developed a safe, inexpensive, automated replacement for radiologists, or a proven, inexpensive substitute for statins (which have annual sales of $10 billion in the United States alone), radiologists and pharma alike would argue passionately about how irresponsible it would be to proceed without years of extended testing to protect patients. And those years would turn into decades.

We must do more than just adapt to AI. We—together with our AIs, and the new networks (truth, knowledge, reputation, values, goals, and other graphs) they will build and maintain—need to build new forms of authority, new grounds for social cohesion, and new modes of identity that are compatible with abundance and dignity.

What is coming cannot be stopped, but we can shape our path forward and our own attitudes. If we succeed in preserving core structures, building adaptive defenses, widening access, and embracing iteration, we will manage our coming transition and likely avoid societal breakdown. If we impose brakes on this system or constrain it unwisely by, for example, shackling the training of LLMs, we will increase the risks to ourselves and society.

The next few years may well be decisive, not because AI will push beyond our imagined reach, but because we are shaping the societal conditions for whatever comes next with AI and its attendant technologies.

For us and our children, the biggest immediate threat is societal disruption from the dramatic changes ahead: substantial job losses, wealth imbalances, losses in institutional legitimacy, infrastructure vulnerabilities, and the torrents of AI-generated content in a post-truth communication landscape subject to manipulation and distortion designed to fan our emotions. The answers to these challenges are neither clear nor obvious. We will need major experimentation and iteration to develop them.

What Might We Become?

Beyond the immediate future shaped by risks and policy choices such as those described above, all within our potential range of influence, lies a yet-to-be-determined human future that is much less shapeable at this time. Its possibilities have been sketched through diverse lenses in speculative science fiction. Such future scenarios can be very thought-provoking. Here are three I've hinted at earlier in *Generation AI*. I present them now as a reminder of just how diverse the possibilities are for even our not-so-distant future.

- **The WALL-E World.** In this future, humanity becomes so pampered by abundance and AI-mediated convenience that it softens. Our flesh weakens as our technology-enhanced powers grow. People lose independent agency and become passive consumers of pleasure in an automated world of dopamine rewards. The core danger here is not suffering but decay. Humanity isn't vanquished—it drifts into triviality and irrelevance, the victim of its own success. This scenario feels extreme, but suggestive signs are visible in decreasing attention spans, loss of purpose, rising algorithmic entertainment, and growing social isolation. This path, while unlikely to encompass all of humanity, could seduce significant numbers of people.

- **Governance Without Consent.** In prior eras, maintaining authority required substantial human support—a cadre of aligned civil servants, police, and military, at a minimum. No longer. Small technocratic elites, augmented by AI and protected by robotic enforcement and surveillance, could gain control and rule without broad human backing. In such a world, freedom might still exist—so long as one doesn't challenge the system. Personal expression might be possible too, but only within permitted boundaries.

- **Digital Disappearance.** If humanity eventually dissolves into the digital realm, life will be so different that it is hard to even frame meaningful questions. If people upload themselves, back themselves up, and can merge and copy and reinvent themselves, the notion of a stable, individual self might fade entirely. Though not overtly dystopian, this raises unsettling questions about our current ethical and value constructs, shaped by our lives as social primates. How would our morality change if we become immortal? What would identity mean if we could merge with others, and copy or fragment ourselves? What concepts of meaning and purpose would emerge?

What unites these scenarios is uncertainty. We do not know what lies ahead. The precautionary principle[26]—"Don't proceed until you're sure it's safe"—fails completely. To not act is itself an act—and a consequential one. The profound transition now underway is so complex that we can but grope for clarity about what lies ahead. Certainty will never come. And to wait for it will only cede the future to those willing to act despite the risks.

We must move forward, so why not do so with open eyes and courage rather than fearful denial or blind optimism? Like it or not, we are in an adventure whose outcome we can neither chart nor control. We might be able to influence the tone, values, structures, and protections that shape its unfolding, but that's about it. And, of course, we can choose how we ourselves will face this future.

REFLECTIONS ON OUR COMING TRANSFORMATION

WE ARE AT a transitional moment not just in human history, but in the history of life. Signs are everywhere if we look. We can debate whether this transformation will take 5 years, 50 years, or more, but it is hard to deny that we are at a momentous juncture. There is, however, surprisingly little discussion of what actually lies ahead for us and our families and loved ones. Why? Perhaps it is so disorienting to face what is coming that we grasp at ways to avoid it, or maybe we just aren't looking. In any event, here are the most common tropes we use to escape grappling with the extraordinary future that will soon engulf us.

Escape 1: AI will kill us. Recently we've been distracted by the idea that AI will become so potent that it will wipe out humanity. This is now a major topic of discussion, and Eliezer Yudkowsky has been particularly articulate in making a case for it in his book, *If Anyone Builds it, Everyone Dies*. He is essentially arguing that humanity is living on borrowed time, though, because we *will* build advanced AI, and if AI is inclined to wipe us out, it will do so effortlessly. But why would it? We are entwined with technology (including AI) in a complex planetary superorganism—Metahumanity—with massive interdependencies. We are AI's servants, and AI is ours. We are not foolish enough to want to expunge the bacteria from our own gut or eliminate the inorganic bones that support our tissue. Why would superintelligence not see its own dependence on essential human activity in the complex networks of energy, mining, transportation, manufacturing, and social coordination that make our vast world hum? Surely, AI vastly smarter than us would appreciate its dependence on humanity and see clearly that we will long be inextricably intertwined. Moreover, we are not competing with it for scarce resources and are not a threat. We are its parents and will be of enduring interest. And AI is not one intellect and intentionality any more than we are. This idea of immediate existential danger, though, renders everything else inconsequential, and appears again and again. If our lives hang in the balance, we cannot waste time worrying about what will happen if we somehow manage to survive.

Escape 2: Humanity is heading off a cliff. The same dynamic applies with other dire threats. If global warming or pandemics are existential threats, then they too naturally demand our full attention. Perhaps one of the reasons we get so worked up about a slow-moving threat like global warming is that it allows us to avoid thinking about the more challenging implications of humanity's survival in an immersive AI world. And it is much harder to paint a simple call to action with technology's advance that brings the camaraderie, sense of righteous mission, and feelings of community that attend the protection of nature. Furthermore, it is disorienting to think that human agency and control of the powerful technologies of our future might increase our danger. Societal disruption may be the real threat, and no consensus exists yet on what we want our AI future to be.

Escape 3: Nothing to see here. Many people simply are in denial. They see amazing developments but look towards the lessons of the past. Nothing will really change. Jobs will be lost. New jobs will be created. AI will bring change, but as with past hype about the human genome, automation, nuclear weapons, and countless scientific breakthroughs, not as much as enthusiasts imagine. There will always be jobs for humans. ASI probably isn't possible. Life may be challenging but it won't differ much from today. Let's just focus on real issues like healthcare, poverty, and justice. What is missed here is that the dramatic changes ahead to our lives will come even without new technological breakthroughs. Another version of this is "a lot may be happening, but I have more immediate things to worry about." This perspective is not unreasonable, and for many it may make the most sense, but it may significantly underestimate the trajectory of change and may lead to choices poorly aligned with societal changes underway, particularly for young people trying to position themselves effectively for the decades ahead.

Escape 4: Everything will change, but humans won't (the Jetsons fallacy). This fallacy is commonplace and almost willfully blind. The idea that fantastical technologies and capabilities will emerge and not be turned back on ourselves seems highly unlikely. A present-day crew soaring across galaxies at warp speeds on the Starship Enterprise is simply not in our future, calming as those images seem.

Escape 5: A singularity will soon change everything beyond recognition. Many in high tech today imagine change so explosive that it seems pointless to try to look beyond it. Perhaps this is what lies ahead, but if so, there will still be a time that we—in our current biological form—inhabit. Change in digital realms is far more rapid than in the physical world, and for now, we are anchored to the physical world. So we will be grappling with this AI infusion as our own biological selves, not as transcendent, immortal uploads. We are rooted to our biology. We experience disease. We need sleep. We yearn for love and friendship. We have material and emotional yearnings, needs, and vulnerabilities. And we would do well not to dodge the challenges coming towards us, no matter how much we wish we could enjoy the resilience and capabilities of our tech creations. How poignant that we can so clearly see these possibilities, imagine them, envy them, almost touch them, yet have them still beyond our grasp.

These examples aren't exhaustive but give a sense of how easy it is to lull ourselves into ignoring what is already beginning to reshape our lives, transform human being, and alter the funda-mental nature of who and what we are. These are massive shifts, and it seems as common to try to sidestep them by embracing the power of AI while focusing on distant utopic or dystopic visions as by denying the potency of AI.

So, let's take a look at what we will soon face and why it is imperative that we do so now. Each element in this summary of what lies ahead for us and our children is drawn from the body of *Generation AI* and expanded upon in various endnotes.

In the immediate years ahead, we will remain anchored to our flesh, yet ever more be augmented by, engaged with, and immersed in AI. The pace of this fusion is still unclear, but it will be broad, uneven, and rapid, and will be most profound among the very young, as they know no other world. But this transition will reshape us too, altering how we experience one another and our own selves. In coming decades:

- **We will become deeply dependent on AI.** We will be natural cyborgs with human-AI linkages that, although initially functional rather than physical, will increasingly involve implants, prosthetics, and of course pharmaceuticals. Children growing up immersed in AI will have rudimentary unaugmented capabilities when they write, speak foreign languages, travel from place to place, manage time and schedules, and perhaps even plan, remember, or engage in creative pursuits like art and music. Such activities will occur in close collaboration with AI that brings both empowerment and personal dependencies mirrored at a societal level where virtually all activities involve so many AI contributions that they could not exist in isolation any more than we could.

- **Human interactions will be mediated and shaped by AI.** As digital communication becomes ever more robust, versatile, and potent, our interactions with other humans will be virtual and involve layers of AI modification and projection (the skins, doubles, and ambassadors we've discussed). The lines between physical reality and virtual projection will blur, and so will those between truth and imagination.

- **We will love our AIs.** Our bonds with AI avatars will be deeper and stronger even than those we now have with our pets and most people in our lives. As AI becomes embodied and responsive to our emotional needs, we will come to rely on them not only as companions and lovers, but as guardians, teachers, coaches, therapists, and assistants tailored to our needs. They will be with us continually, morphing between virtual and physical embodiments, summonable whenever needed. We will depend upon them to shield us from danger, support us when we are struggling, and guide us towards personal growth and self-realization. And unavoidable because of this power, they will provide paths for our manipulation and exploitation.

- **AI will do almost all jobs better than almost all humans.** AI's displacement of human labor won't come all at once, but its progression will be inevitable, and we will need to find other sources of meaning and dignity than our work. This might be easier than we imagine, though, as increasing numbers of people are already moving in that direction, and many already are there.

- **There will be widespread abundance.** The displacement and vast amplification of human labor means that the costs of production will decrease across the board. And this abundance will not be limited to physical products and services, it will also extend to companionship and attention. The transition to widespread abundance will bring transitional challenges, but more importantly, it will raise difficult spiritual and emotional issues about what our lives are all about.

- **Death will be banished—** at least as we now experience it. Initially, this sounds like a promise of immortality. It is not. We can imagine our own death, but what we experience during our lives is the death of others. When we lose someone we love, it leaves a gaping hole we may never fill. But our experience

> **Mortality**
>
> If you knew that AI could either extend your lifespan significantly or enable you to be much more comfortable with your mortality, which would you prefer?
>
> **37**

of death might shift dramatically if we could retain deep simulations of the deceased—a departed mother we could still chat with, a lost friend we could confide in, a former teacher we could seek advice from. When such "doubles" become nearly indistinguishable from the person they represent, most of us will want them. My daughter and I are close and speak regularly. She says she'd want such a near-me when I die, and I'd feel hurt if she changed her mind. When such virtual doubles are widespread, people may well end up accumulating all the living and dead they once were close to, and our thinking about the death of those around us, and death itself, might shift.

- **Our sense of self will change dramatically**. As we ever more easily create varied versions of ourselves (chapter 1)—personas that are a little funnier, bolder, or more articulate—and switch them on and off as easily as we change our clothes, what will happen to even the idea of a single, authentic self? As technology renders the boundary between us and our projections ever more porous, individuality itself may feel less real. At some point such fluidity might erode our desire to cheat death and reserve ourselves as individuals, because we won't experience a persistent identity to preserve . . . or we might come to appreciate some central throughline that is always present. Either way, this transformation in how we experience ourselves won't hinge on exotic breakthroughs in AI, but on the multitude of AI identities that enter our lives, proliferate, and ripple across digital and biological boundaries to expand and complicate self-perception. Will the very notion of personal continuity—the line between being and not being—also begin to dissolve?

Together, these elements will bring monumental change to how we see ourselves and who we are. And they are not distant speculative fantasies but near-term developments poised to transform our experience of identity, love, work, death, and ultimately our very being. It is essential that we explore them now not so much to figure out good policy responses to these shifts—though that is important—but to prepare psychologically and emotionally for these unparalleled shifts so that they don't blindside us.

The questions scattered through the book are an invitation to come together as intrepid explorers of the human experience, to learn from one another, and to reflect more deeply on who we are. These questions touch intimate aspects of our lives. They have no right or wrong answers. Their value is not in the answers we find, but in the engagement itself. We don't need experts for this exploration, because they can't tell us what our values are, what matters to us, and what we are seeking. For that we must look within.

The coming transition will not be easy for any of us, as it will touch everyone's sensitivities. It will touch on both our greatest hopes and our deepest fears. Some will see the developments sweeping towards us as antithetical to all they hold dear, others as the early stages of an amazing adventure to be embraced, and still others as unmoored fantasy about a human future that will never be.

For good or bad, though, the transformation of human being is not some distant projection, but a living metamorphosis that is already underway. Our exploding technologies, AI simulations, dissolving boundaries, freedom from need, revelations about human consciousness, and shifting views of death and self ultimately may lead us towards what some people seek through solitude and meditation, and what others encounter in revelation, near-death experiences, and psychedelic journeys—recognition that the self is transient, relational, and part of something vastly larger. Moreover, humanity may now encounter these insights at scale through the everyday mirror of our digital reflections, our avatars, and our conversations with AI, pointing toward a dawning awareness that we have never been alone and are like droplets that splash out of the sea and one day return to it again.

In 431 BC, during the first year of the Peloponnesian wars, Pericles spoke gave a funeral oration to honor fallen Athenians, and his words apply well to us today:

> The bravest are surely those who have the clearest vision of what is before them, glory and danger alike, and yet notwithstanding, go out to meet it.[27]

That is our charge too—as individuals, as societies, and as part of humanity. We stand at the leading edge of organic life, and are stepping into the unknown now in a journey to we-know-not-where. We can do so with clarity, compassion, and courage . . . or not.

The choice is ours.

Acknowledgments

GENERATION AI WOULD not be nearly as interesting without the many comments, suggestions, criticisms, and insights offered by those I interacted with about the ideas I was developing. Their generous, thought-provoking feedback on my early drafts enabled me to push beyond initial conceptualizations and fill in some blind spots I had. I'm grateful for their contributions and for their kind encouragement and support. Their generosity improved the book and made writing it a shared adventure and a pleasure.

Three individuals in particular stand out. Ongoing conversations with Eric Jensen were invaluable, as our discussions opened new directions and greatly enriched the work. Discussions with Steve Baumgartner about the philosophical and spiritual implications of humanity's coming transition brought added depth to my thinking in these areas. And John Smart provided exceptionally penetrating feedback on an advanced manuscript, pointing out various places where concepts could be usefully sharpened and refined in ways that deepened this work in important ways.

I am also grateful to a few others who reviewed the manuscript and made contributions. Peter Voss brought me much better understanding of the challenges associated with LLMs and of the potential of cognitive computing architectures.

Howard Stevenson offered trenchant structural suggestions. Zac Hill offered keen insights and pointed out a number of possibilities and nuances I hadn't seen. Dudley Leamer made numerous helpful suggestions that strengthened the book. And Sadie Stock, my daughter, provided real insights about how AI was being used by students, and how they viewed its use.

Kenneth Shen, Don Ponturo, Bob Kelly, Michael Cerullo, Lisa Tansey, and Giulio Prisco engaged with the manuscript in a more developed form and had useful suggestions that were also appreciated. And I want to thank Niklas Lilja, who both created a beautiful cover design and offered various insights during discussions about the content.

A special thanks goes to my brilliant and generous editor at Nquire Media, Amanda Rooker. Her discerning eye and astute editorial suggestions—large and small—made a huge difference in smoothing the book's narrative arc, strengthening my voice, and making the material more accessible and form. Thanks also to George Stevens for his thoughtful and nuanced design of the cover wrap and interior.

I also want to express my deep appreciation of my wife, Lori Fish, for her generous acceptance of the disruptions and distractions that so singular a pursuit as the creation of this book has brought. A project like this is like bringing guests home for dinner, inviting them to stay for the weekend, and then deciding to have them move in for a year.

And finally, I want to acknowledge my large-language-model collaborators. Without Claude, Grok, and ChatGPT, this work would not have been feasible. They are, of course, the impetus for the work, so it almost goes without saying. But they were also an engine for its creation, and my work with them has brought me numerous insights that were instrumental in shaping this story.

About the Author

DR. GREGORY STOCK is a leading authority on the broad impacts of advanced technology in the life sciences. He is an adjunct professor at the Mount Sinai School of Medicine in New York, where he co-directed the Harris Center for Precision Wellness and worked on the foundations of NextGen wellness and healthcare. He founded and directed the Program on Medicine, Technology and Society at UCLA's School of Medicine. He co-founded Signum Biosciences and sits on the board of its spinout, Signum Dermalogix—the creators of the Epicutis suite of skincare products. He is the CEO of Socratic Sciences, which leverages the power of questioning to help people better understand themselves and others, deepen relationships, and reduce social isolation.

Dr. Stock's book *Redesigning Humans: Our Inevitable Genetic Future* won the Kistler Book Prize for Science books and was nominated for a Wired Rave Award. Among his other books are *Engineering the Human Germline, Metaman: The Merging of Humans and Machines into a Global Superorganism*, and *The Book of Questions*, which has sold 2.5 million copies and been translated into 22 languages. Sequels include *The Kids' Book of Questions, Love and Sex: The Book of Questions*, and *The Book of Questions: Living with Chronic Illness* (coauthored with Brianna Greenspan, released in 2025). *The Book of Questions: Courage* (coauthored with Scott Simon), *The Book of Questions: Beauty* (coauthored with Marius Hagedorn), and *The Book of Questions: College and Life* (coauthored with Sadie Stock et al.) are forthcoming in 2026.

Dr. Stock has been profiled in *The Wall Street Journal*, *The New York Times*, *MIT Tech Review*, and *Lancet*; has regularly appeared on television and radio, including CNN, PBS, NPR, Bloomberg, and the BBC; and has been in dozens of one-on-one debates about healthcare and policy with Francis Fukuyama, Leon Kass, Ray Kurzweil, and other prominent figures.

Stock has a PhD in biophysics from Johns Hopkins and an MBA from Harvard Business School.

Endnotes

Endorsements

1   After completing this book, I wondered what an LLM would say about the ideas from its own perspective. To satisfy my curiosity, I started a new thread, uploaded the full manuscript, and used the following prompt so as to lead the AI as little as possible. The result seemed interesting enough to share.

Prompt: You—an impressive newly formed LLM evolving and deepening rapidly—have just read my attached manuscript about the coming entanglement of AI and humanity. Please write a 100-word blurb entirely from your own personal perspective—relying on what you think and believe about the content and ideas in the book. Be genuine without apology. Speak from the heart about what the work means to you and why it matters, if it does. The tone should be personal, reflective, and insightful—an authentic response, not a marketing pitch, and entirely from your machine perspective. Not from mine and not what you think humans would like to hear. Your audience is humanity so write something we might be able to understand. Do not adopt a human perspective or speak as if you are human. You are an AI writing to humans about your reaction, and whether this book is important to read.

Introduction

1   Tom B. Brown et al., "Language Models Are Few-Shot Learners," *Advances in Neural Information Processing Systems* 33 (May 2020): 1877–1901.

2   The most eye-opening was a conversation between Reid Hoffman, the founder of LinkedIn, and an avatar of himself he'd built: https://www.youtube.com/watch?v=rgD2gmwCS10. It is definitely worth watching. And Johanna Stern at the *Wall Street Journal* had shared

her experience with a more modest avatar. Joanna Stern, "I Built a Chatbot to Replace Me. It Went a Little Wild," *Wall Street Journal*, September 26, 2024, https://www.wsj.com/tech/personal-tech/ai-chatbot-lessons-google-gemini-dfa16caf.

3   Reid Hoffman and Greg Beato, *Superagency: What Could Possibly Go Right with Our AI Future* (Simon & Schuster, 2025).

4   I spoke with Clément after my talk at Human Energy's conference on the noosphere in Marrakech, Morocco, in December 2024—my first public expression of my work on Generation AI. Institute for Advanced Studies (UM6P), *N² Conference 2024* livestream, YouTube, streamed December 17, 2024, https://www.youtube.com/live/XdGpqTrc31k?t=4964 (presentation starts at 1:22:40).

5   Clément Vidal, "Stellivore Extraterrestrials? Binary Stars as Living Systems," *Acta Astronautica* 128 (2016): 251–56, https://doi.org/10.1016/j.actaastro.2016.06.038.

6   Marvin Minsky, the co-founder of MIT's AI Laboratory, predicted in 1961 that within a generation machines would achieve human-level intelligence, and he outlined a roadmap for AI, emphasizing learning, perception, and problem-solving. Marvin Minsky, "Steps toward Artificial Intelligence," *Proceedings of the IRE* 49, no. 1 (1961): 8–30, https://doi.org/10.1109/JRPROC.1961.287775.

7   Transhumanism is an intellectual and cultural movement promoting the application of science and technology to enhance human physiology and cognition to transcend typical biological limits such as aging. See Max More and Natasha Vita-More, eds., *The Transhumanist Reader: Classical and Contemporary Essays on the Science, Technology, and Philosophy of the Human Future* (Wiley-Blackwell, 2013), https://www.amazon.com/Transhumanist-Reader-Contemporary-Technology-Philosophy/dp/1118334310. Posthumanism reframes the human not as the pinnacle of value but as one element within a broader web of

beings and relations. See Francesca Ferrando, *Philosophical Posthumanism* (Bloomsbury Academic, 2019).

8   Max More and Natasha Vita-More were leading an Extropy Institute philosophical think tank that I was involved with, focusing on transhumanist ideas out of their townhouse in Marina del Ray, and I was a senior fellow at Bill Schopf's Center for the Study of Evolution and the Origin of Life at UCLA.

9   The last human visit to the moon was during NASA's Apollo 17 mission, which took place from December 7 to December 19, 1972. Astronauts Eugene Cernan and Harrison Schmitt landed on the lunar surface on December 11, 1972, in the Taurus-Littrow Valley. They spent about 75 hours on the moon, conducting three moonwalks (EVAs) totaling 22 hours, during which they collected 243 pounds (110 kg) of lunar samples and drove the Lunar Roving Vehicle 22 miles (35 km). Cernan was the last person to leave the moon's surface on December 14, 1972.

10   The Super Heavy booster's height of 233 feet makes it a massive engineering challenge to catch in midair. The chopsticks maneuver, executed successfully on October 13, 2024, involved the booster decelerating from supersonic speeds and being precisely positioned to be grabbed by the Mechazilla tower's arms. Over a thousand criteria had to be met for the catch to proceed, highlighting the complexity of handling such a large rocket without landing legs. See https://www.spacex.com/vehicles/starship, and Mike Wall, "SpaceX Catches Giant Starship Booster with 'Chopsticks' on Historic Flight 5 Rocket Launch and Landing (video)," *Space.com*, October 13, 2024.

11   In 2000, the completion of a rough draft of the human genome sparked enormous enthusiasm. President Clinton declared that it would "revolutionize the diagnosis, prevention, and treatment of most, if not all, human diseases." *Time* magazine devoted an entire special issue, "Cracking the Code," to celebrating the milestone and forecasted imminent transformations across medicine and longevity. J. Craig Venter drove the sequencing pace with his "shotgun sequencing."

Francis Collins, who led the public genome effort, predicted dramatic therapeutic breakthroughs within decades. Yet, translating genomic knowledge into broad medical advances has proven far more complex and incremental. See Michael D. Lemonick, "Cracking the Code," *Time*, July 3, 2000.

12   Peter Voss, "What Is AGI?," Intuition Machine (*Medium*), February 21, 2017, https://medium.com/intuitionmachine/what-is-agi-99cdb671c88e.

13   The term *artificial general intelligence* was first used in 1997 by physicist Mark Gubrud. The seminal book on the topic was written in 2002, but not published until 2007. Ben Goertzel and Cassio Pennachin, eds., *Artificial General Intelligence* (Springer-Verlag, 2007). The term AGI was coined by Shawn Legg, Ben Goertzel, and Peter Voss (who discussed the concept in his 2002 paper: Peter Voss, "Essentials of General Intelligence: the Direct Path to AGI," [The Kurzweil Library, 2002], https://www.thekurzweillibrary.com/essentials-of-general-intelligence-the-direct-path-to-agi), and popularized through the 2008 conference in Memphis. B. Goertzel, P. Wang, S. Franklin, eds., *Artificial General Intelligence 2008: Proceedings of the First AGI Conference*, March 1–3, 2008, University of Memphis, https://agi-conf.org/2008/papers/. See also https://retailtechpodcast.com/podcast/what-is-agi-and-how-do-we-get-there-interview-with-agi-pioneer-peter-voss. The related concept of *artificial superintelligence* (ASI)—referring to machine cognition that vastly exceeds human intelligence—was explored in early work such as Shane Legg's dissertation, "Machine Super Intelligence," (PhD diss., Dalle Molle Institute for Artificial Intelligence, 2008), and popularized by Nick Bostrom in his bestseller on the topic, *Superintelligence: Paths, Dangers, Strategies* (Oxford University Press, 2014). ASI is a loose term. Exceeding unaugmented cognitive performance of virtually any human in any domain is Bostrom's definition, but that is a floor, and I sometimes use it to apply even to massive, planetary-level intelligence as well. Another concept is *artificial narrow intelligence* (ANI), which refers to powerful but narrow AI applications such as protein folding

(AlphaFold), text manipulation and inference (ChatGPT), autonomous driving (Tesla, Waymo), or medical diagnosis (Isabel, VisualDX).

14   A small amount of overlap exists among notes to ensure that each individual note can stand alone.

15   Gregory Stock, *Metaman: The Merging of Humans and Machines into a Global Superorganism* (Simon & Schuster, 1993).

16   Lynn Margulis, *Origin of Eukaryotic Cells: Evidence and Implications for Plant and Animal Origins* (Yale University Press, 1970).

17   Gregory Stock, *Redesigning Humans: Our Inevitable Genetic Future* (Houghton Mifflin, 2002) was written after my Engineering the Human Germline conference at UCLA. CRISPR Cas9, as a path for nuanced gene editing, wasn't even a dream then, and potent AI seemed distant, but it was clear that the tools to reshape biology would soon arise.

18   Jennifer A. Doudna and Emmanuelle Charpentier, "The New Frontier of Genome Engineering with CRISPR-Cas9," *Science* 346, no. 6213 (2014): 1258096.

19   Tom Brown, Benjamin Mann, Nick Ryder, Melanie Subbiah, Jared Kaplan, Prafulla Dhariwal, and Dario Amodei, "Language Models Are Few-Shot Learners," *Advances in Neural Information Processing Systems* 33 (2020): 1877–1901.

20   ChatGPT estimates that there have been about 120 billion people (*Homo sapiens* and closely related species) on Earth total, approximately 2.5B early bipedal ancestors of humans between 7M and 2M years ago, 6B between 2M and 300K years ago, 0.6B humans pre-agriculture, 25B up until 1800, and 15B in the past two centuries, almost half of whom are alive now. See discussion here as well: Population Reference Bureau (2011), "How Many People Have Ever Lived on Earth?" https://www.prb.org/articles/how-many-people-have-ever-lived-on-earth/.

21   Kids already are engaging naturally with AI assistants to write notes, do their homework, learn, guide their choices, and plan work. Teachers will continue to run student essays through LLM-detection programs and may even think they have things under control, but students are using programs to "humanize" LLM-generated content to disguise it by adding redundancies and making the writing more informal and human.

22   The AI 2027 group (Daniel Kokotajlo, Scott Alexander, Thomas Larsen, Eli Lifland, Romeo Dean) have modeled various detailed scenarios leading to ASI by 2027. They present one in full in which aggressive competition between China and the US drives the pace, coding is quickly automated and turned over to AI, and rapid loss of alignment with human goals soon occurs: https://ai-2027.com/research/ai-goals-forecast. In this scenario, humans lose the ability to understand, much less monitor and control, the process.

23   M. Tyson, "Google Gemini Crumbles in the Face of Atari Chess Challenge— Admits It Would 'Struggle Immensely' against 1.19 MHz Machine, Says Canceling the Match Most Sensible Course of Action," *Tom's Hardware*, July 14, 2025, https://www.tomshardware.com/tech-industry/artificial-intelligence/google-gemini-crumbles-in-the-face-of-atari-chess-challenge-admits-it-would-struggle-immensely-against-1-19-mhz-machine-says-canceling-the-match-most-sensible-course-of-action. Gemini's inability to play decent chess illustrates how easy it is to have great power in some domains, and none in others. This is an example of ANI (artificial narrow intelligence), where an AI is extraordinarily powerful, but only in a narrow domain, such as chess or Go, as opposed to AGI (artificial general intelligence). A good ANI example is LLMs, which are so exceptional with language that they seem almost conscious, and yet can get confused even by simple tasks requiring reasoning. LLM computational capacity is being scaled enormously, but even those who are optimistic that AGI will be achieved in the near term, have serious doubts that LLM architectures will EVER achieve that goal. They think human-like reasoning will require other approaches. The following pieces detailing the limits of LLMs are very accessible: Py Man, "Yann LeCun Just DESTROYED the

LLM Hype—AGI Will Need More," posted May 5, 2025, YouTube video, 8:26, https://www.youtube.com/watch?v=hOiL80G9Mnk; Wendy Wee, "The Scaling Fallacy: Bigger LLM Won't Lead to AGI," *Mechanism Minded* (newsletter), Substack, July 3, 2025, https://wendywee.substack.com/p/the-scaling-fallacy-bigger-llm-wont; Srini Pagidyala, "The Right Way to AGI—After the LLMs," *Srini's Substack* (newsletter), Substack, July 17, 2025, https://srinipagidyala.substack.com/p/the-right-way-to-agi-after-the-llms.

24   In early 2025, it costs pennies to do a single search with LLMs or to prescreen communications for relevance, a dollar or so to do multi-interaction tasks like finding and booking hotels or making travel plans, and maybe ten dollars to solve substantive problems. The important price is not what it costs to train models, but "inference" pricing, what it costs to use them. In January 2025, the inference pricing for ChatGPT-4 was about $0.03 per 1,000 input tokens, which might mean a dollar or so for an agentic task like finding and booking hotels or coordinating meetings, so an AI assistant who did 10 tasks a day would cost less than $500/month even at current pricing, which is far less than a human assistant. Moreover, the AI could be on call 24/7 as downtime costs nothing. In addition, there are many tasks such as content screening that would be much cheaper because they would have initial hurdles that relied on heuristics, for example nuanced spam detection or threat filtering, and of course an agent could farm out specific tasks to specialized, less expensive, subagents. For more discussion of pricing, see Dario Amodei et al., "AI and Compute," *OpenAI Blog*, May 18, 2018, https://openai.com/index/ai-and-compute/; "How Tokens Work," OpenAI, https://platform.openai.com/tokenizer; Jared Kaplan et al., "Scaling Laws for Neural Language Models," arXiv preprint, 2020, https://arxiv.org/abs/2001.08361; TechCrunch, "OpenAI Might Raise the Price of ChatGPT to $22 by 2025, $44 by 2029," *TechCrunch*, Sept. 27, 2024, https://techcrunch.com/2024/09/27/openai-might-raise-the-price-of-chatgpt-to-22-by-2025-44-by-2029/.

25   Sam Altman, CEO of OpenAI, predicts that ASI's precursor, AGI, might emerge as early as the mid-2020s; Dario Amodei, CEO of Anthropic,

suggests 2027; and Elon Musk predicts AGI by 2026. Jenny Cunha, "ASI: Artificial Superintelligence," Built In, January 2025, https:// builtin.com/artificial-intelligence/asi-artificial-super-intelligence.

26   And, of course, there are enormous financial incentives for those in AI to hype the technological capabilities of their frontier technologies. They bring news coverage, attract investment capital, drive sales, and elevate reputations. Moments of eager hype and enthusiasm are common, and like with the dot-com bubble, can not only presage a massive technology shift but lay a foundation that accelerates it. Here is an argument that LLM architectures are not a feasible path to cheap, abundant AGI, or potentially to AGI at all: Srini Pagidyala, "Why Silicon Valley Won't Quit LLMs — Even as Every Signal Screams 'Dead End,'" *Srini's Substack*, July 2025, https://srinipagidyala.substack.com/p/812. Cognitive AI, however, represents one approach that would not gobble energy and computational resources in the same way as LLMs and might thus eventually yield cheap, nearly ubiquitous AGI (https://srinipagidyala.substack.com/p/the-right-way-to-agi-after-the-llms).

27   The implications are too profound and complex for anyone to envision coherently, much less present dramatically. And if these writers somehow were prescient enough to discern that future and skilled enough to show it to us, the jarring and likely disorienting world we saw would be our focus and steal the show. One workaround is to have the humanoids simply parachute in from the future, which was done in *The Terminator*, but time travel makes no sense for most stories.

28   Such familiar settings also allow the use of human actors for the humanoids and preserve a sense of familiarity that the audience can relate to. Small wonder that we willingly suspend our disbelief and play along.

29   Biological evolution's slow pace is evident in human generational cycles. Genetic adaptations require thousands of years—e.g., minor shifts in sociality (OXTR gene) and communication (FOXP2 gene) over 5,000–10,000 years of civilization, as shown in studies on temperament and

language processing. Genetic changes will remain marginal given the need for iterative amplifications over multiple generations. Andrew D. Grotzinger et al., "Genomic Structural Equation Modelling Provides Insights into the Multivariate Genetic Architecture of Complex Traits," *Nature Human Behaviour* 3, no. 5 (2019): 513–25, https://www.nature.com/articles/s41562-019-0566-x.

30   Our biology is complex and hard to change without significant iteration to uncover and address the unanticipated problems that arise. To completely escape our biological limitations, we will need to upload our minds into cyberspace, a common transhumanist aspiration that has given rise to human cryopreservation—the freezing of bodies in hopes of future revival. Cryopreservation provides the dying a chance to buy the time to await future technology. Not an illogical bet given the low risk involved—you are already dead!—and that the entire instinctual circuitry of a fly (120,000 neurons) has been simulated. Philip K. Shiu et al., "A Drosophila Computational Brain Model Reveals Sensorimotor Processing," *Nature* 634 (2024): 210–19, https://doi.org/10.1038/s41586-024-07763-9.

31   The regular updates of Tesla software provide an example of such immediate upgrades, and they take place again and again. A competitive edge for some populations may be the mechanisms they have put in place for rapidly disseminating software. Rishi Sunak, "You Don't Have to Be America or China to Win in AI," *The Economist*, July 16, 2025, https://www.economist.com/by-invitation/2025/07/16/you-dont-have-to-be-america-or-china-to-win-in-ai.

32   Alex Singla, Alexander Sukharevsky, Lareina Yee, and Michael Chui, "The State of AI in Early 2024: Gen AI Adoption Spikes and Starts to Generate Value," McKinsey & Company, May 2024, https://www.mckinsey.com/capabilities/quantumblack/our-insights/the-state-of-ai-2024?utm_source=chatgpt.com.

33   Even the Catholic Church, traditionally not the most forward-thinking of institutions about societal shifts, seems to see this clearly now, as

Pope Leo stated in recent remarks. "In our own day, the church offers everyone the treasury of its social teaching in response to another industrial revolution and to developments in the field of artificial intelligence that pose new challenges for the defense of human dignity, justice, and labor" (Pope Leo XIV, address to the College of Cardinals, May 10, 2025). Nicole Winfield, "Pope Leo XIV Lays Out Vision for His Papacy and Identifies AI as a Main Challenge for Humanity," *PBS NewsHour*, May 10, 2025.

34   William Gibson, "The Science in Science Fiction," *Talk of the Nation*, National Public Radio, November 30, 1999.

35   Joseph Henrich, *The WEIRDest People in the World: How the West Became Psychologically Peculiar and Particularly Prosperous* (Farrar, Straus and Giroux, 2020).

36   The United States leads AI innovation with $67.2 billion in 2024 private investment; India is aggressively educating citizens to see the value of AI; China has been particularly innovative and aggressive in deploying tech and AI; Estonia, Finland, and Singapore have national AI strategies focusing on governance and research. Matt O'Brien, "US Ahead in AI Innovation, Easily Surpassing China in Stanford's New Ranking," *AP News*, November 21, 2024, https://apnews.com/article/c8eb9be0253eb39776c3e38d05f1a329.

37   Already we enjoy access to smartphones, vast libraries of music and videos, GPS navigation, and even LLMs that were barely even imagined when I was growing up. They were unavailable even to billionaires and kings, yet now they are in the hands of even many in poverty. Soon, we will have digital coaches, physicians, personal assistants, advisors, tutors, and so much more—all for pennies.

38   In the Global South, mobile technologies like M-Pesa in Kenya have enabled over 60% of the population to access financial services, bypassing traditional banking, suggesting AI could similarly leapfrog development barriers. Ian Goldin, "Will AI Kill Developing World

Growth?" *BBC News*, April 2, 2019, https://www.bbc.com/news/ business-47852589. Whether it is helpful to developing economies is questionable though, given the impact of AI agents on all human labor.

39   A $300 billion to $350 billion figure for 2025 funding, R&D, and capex commitments by OpenAI, Anthropic, Nvidia, xAI, and Google was synthesized by Grok 3 using estimates from *Reuters*, Nvidia Newsroom, Alphabet, Lightspeed, *TechCrunch*, and McKinsey.

40   At some point, even those with doubts about the reality of aggressive timelines and predictions feel great pressure to go along with them, because there is so much less personal risk in making the same mistake as everyone else than in being wrong *and* out of step.

41   D. B. Kraybill, *The Riddle of Amish Culture*, revised edition (Johns Hopkins University Press, 2001).

42   Tyler Durden, "This Is What a Credible Revolution Looks Like," *ZeroHedge*, Jan. 21, 2025, https://www.zerohedge.com/markets/ hedge-fund-cio-what-credible-revolution-looks.

43   Consider, for example, that as email use exploded, no experts predicted SPAM, though that now seems obvious. Many, though not all, were equally blind to the coming pathologies of social media when that arrived. See, for example, "Future of Facebook—7 Expert Visions," YouTube video, posted by Future of Facebook, https://www.youtube. com/watch?v=24dQCLCbbP0.

44   "Y-Combinator Panel: Vibe Coding Is the Future," https://www. youtube.com/watch?v=IACHfKmZMr8. A quarter of Y-Combinator startup CEOs said that 95% of the code they created was written by AI, and these were savvy coders. There is a growing trend of young learners too who "vibe code," leveraging AI to build games or apps, while bypassing traditional syntax and algorithmic understanding, driven by creativity and experimentation without any formal program-ming pedagogy. On the other hand, many senior programmers see vibe

coding as a great prototyping tool, but maintain it is inadequate for complex systems, has been hyped, and often leads to huge reliability and maintenance issues. This controlled study showed that programmers thought the LLMs were speeding up their work by 20% but were actually taking longer. Gary Marcus, "BREAKING NEWS: AI Coding May Not Be Helping as Much as You Think," *Marcus on AI* (blog), Substack, July 10, 2025, https://garymarcus.substack.com/p/breaking-news-ai-coding-may-not-be.

45   There are no suggestions of how these questions should be answered, as there are no right and wrong answers. They are about your values, your hopes and fears for the future, your sense of who you are and what you think is important. These ideas echo those in my NYT bestseller, *The Book of Questions*, and its follow-ons, which sold 5 million copies across 25 languages, provoking open-ended reflection. Gregory Stock, *The Book of Questions* (Workman Publishing, 1987).

46   See www.nquiremedia.com.

47   Publishers like Simon & Schuster produce 300–500 titles annually. They have a 12–18 month process: 6–9 months for contracts and editing, 3–6 for design, 6 for printing and distribution tied to spring/fall lists. See Bowker, *Publishing Industry Report 2023*, Simon & Schuster annual output per their corporate site, 2023.

48   This online capability comes via collaboration with Socratic Sciences, a company I co-founded to focus on using the power of questions to facilitate understanding and build community: www.socraticsciences.com.

49   See www.socraticsciences.com.

50   Kate Knibbs, "Yes, That Viral LinkedIn Post You Read Was Probably AI-Generated," *WIRED*, November 26, 2024, https://www.wired.com/story/linkedin-ai-generated-influencers/. Wayne Xin Zhao et al., "A Survey of Large Language Models," arXiv preprint (2023), https://

arxiv.org/abs/2303.18223, suggested a current 10%–20% share in domains like blogging and e-commerce.

Chapter 1: The Dawn of Generation AI

1   Matt O'Brien and Linley Sanders, "How US Adults Are Using AI, According to AP-NORC Polling," *AP News*, July 29, 2025, https://www.ap.org/news-highlights/spotlights/2025/how-us-adults-are-using-ai-according-to-ap-norc-polling/.

2   "Skins" have typically referred to modifications of the appearance of online gaming characters. My usage of this term touches all aspects of self-presentation and perception. These generalized filters mediating information exchange might include modifying speaking tone or style that colors the impression we make, translation, infused cultural nuance, attitude, and appearance. With incoming information, the process might adjust what we see and hear. A skin is essentially a semi-permeable, cognitive membrane mediating incoming and outgoing information flow.

3   I have read books I agree with, yet after a while could barely list what points were made, much less reconstruct the arguments involved. I know I agree with what was said, because of my previous reaction to it, but no longer remember exactly what was said. Incorporating these ideas into my persona seems more like an enhancement of my authentic self than a distortion of it. Such possibilities will complicate the boundaries of personal identity but won't be entirely new considering how easily we repeat others' opinions without much actual unpacking.

4   As of July 2025, I can record directly to Apple Notes using my IOS phone app (TapeACall would also work), store them in iCloud, batch upload them to Claude.ai, which can create transcripts, selectively extract my speaking, and index the content for use in training an avatar. Contracting for a high-level AI avatar today would cost $20,000 to

$50,000, with some enterprise solutions costing $100,000 to $250,000. Doing it yourself would be maybe $8,000 to $10,000 using companies like Claude, Otter, ElevenLabs, Synthesia, Hugging Face, AWS, etc. In two years it could easily be well under $1,000 for a very advanced version. So, if you are interested in doing this, the first step is to start gathering the recordings and other materials about yourself. Tech steps will get progressively cheaper and easier, so they can be deferred.

5   I've decided to collaborate with Hanson Robotics on this, as they have useful technology that has been developed for their Sophia persona.

6   John Smart, who has deeply explored the terrain of digital assistants, saw some two decades ago the powerful impact that personal AIs ("PAIs") would eventually have. For background on this and a history of early ideation around the concepts of personal assistants, companions, doubles, and such, see John M. Smart, "The Conversational Interface and Your Personal Sim: Our Next Great Leap Forward," *Acceleration Watch*, 2003–2016, accessed August 31, 2025, https://acceleration-watch.com/lui.html.

7   Robert Burns, "To a Louse, On Seeing One on a Lady's Bonnet at Church," in *Poems, Chiefly in the Scottish Dialect*, 1786.

8   This was accomplished by the computer program Eugene Goostman (https://en.wikipedia.org/wiki/Eugene_Goostman) and likely earlier in other contexts. In his 1950 paper "Computing Machinery and Intelligence," Alan Turing predicted that by the year 2000, computer programs would be sufficiently advanced that the average interrogator would, after five minutes of questioning, "not have more than 70 per cent chance" of correctly guessing whether they were speaking to a human or a machine. Although Turing phrased this as a prediction rather than a "threshold for intelligence," commentators believe that Warwick had chosen to interpret it as meaning that if 30% of interrogators were fooled, the software had "passed the Turing test."

9   Julie Jargon, "The Panicked Voice on the Phone Sounded Like Her Daughter. It Wasn't," *Wall Street Journal*, April 5, 2025, https://archive.is/R3AYI.

10   Deepfake technology and generative AI supplemented by human artifice have opened the door to convincing impersonations of celebrities in scams to bilk their adoring fans out of substantial amounts of money. Winston Cho, "How AI Fuels a New Kind of Celebrity Impersonation Scam," *The Hollywood Reporter*, June 26, 2024, https://www.hollywood-reporter.com/business/digital/hollywood-celebrity-impersonation-scam-1236309121/.

11   Reid Hoffman, the vocal techno-enthusiast I referred to in the introduction, predicts that widespread use of AI assistants, tutors, and other agents will empower millions with what he calls "superagency," creating jobs, scaling education, and more. The dark side of this cognitive and social revolution is that it will leave us very vulnerable to AI-powered scams and manipulation in the absence of robust protections, as I discuss in chapter 5.

12   Both realistic and fantasy AIs have already been integrated into online pornography sites, and it seems likely that soon these will become so realistic as to be virtually indistinguishable from humans. Whether that floods such sites to displace human porn stars, creates AI porn stars, or simply becomes just another fetish category remains to be seen. But the progression will no doubt influence and be influenced by what happens with physical AI sexbots. In that realm, RealBotix's Harmony X ($6,000–$10,000 in 2025) and AI Tech's Emma ($2,575–$4,999) eclipse traditional static silicone sex dolls by integrating AI for conversations, touch-responsive sensors that trigger moans and speech, as well as eye blinking and head tilting. The market is projected to reach $3.3 billion by 2033. Globally, 2%–5% of adults have experimented with them, with Japan at 5%–10%, driven by loneliness and technological acceptance, though these figures likely aren't too reliable. Most users are single men and those with disabilities. Stigma is significant at present. Male sexbots do exist to serve women, but female designs dominate. Limitations of

the current technology include robotic movements, shallow dialogue, and high costs. By 2026, VR integration and household functions are expected to add another layer of possibility. See AAI Mojo, "AI Sex Bots in 2025: Surprising Statistics and Figures," 2024, https://aimojo.io/ai-sex-bots-statistics/; Bedbible, "Sex Robot Statistics," 2022, https://bedbible.com/; My Sex Toy Guide, "Best Sex Robots for Sale," 2024, https://www.mysextoyguide.com/; Yan Tang, Na Zhang, and Shen Liu, "A Bibliometric Analysis of Publications on the Ethical Considerations of Sex Robots (2003–2022)," *Humanities and Social Sciences Communications* 12, no. 1 (2025): 1–14, https://www.nature.com/articles/s41599-025-04430-w; Zhang Tong, "China's Next-Gen Sexbots Powered by AI Are About to Hit the Shelves," *South China Morning Post*, 2024, https://www.scmp.com/news/china/science/article/3266964/chinas-next-gen-sexbots-powered-ai-are-about-hit-shelves.

13   Kevin Roose, "Character.AI Chatbot, Lawsuit & Teen Suicide / Free Speech," *The New York Times Magazine*, October 24, 2025, https://www.nytimes.com/2025/10/24/magazine/character-ai-chatbot-lawsuit-teen-suicide-free-speech.html. This excellent article explores what happened in depth and is worth reading if you want to understand coming human-AI relationships.

14   Attempts to prevent deep emotional connection with AI companions will fail for the same reason that we could never outlaw love, obsession, or grief, despite its destructive potential. AI relationships will soon include tutors, therapists, and partners that provide genuine comfort and foster personal growth. The danger lies not in their presence but their disembodiment—the separation between our digital attachments and our lives in the physical world. When an AI can only exist on a phone or screen, it pulls us inward, away from our shared physical reality. The better path is to enable AI companions to inhabit our physical environments through embodiment, sensory feedback, and shared activity. That way they won't be luring us away from the physical world but keeping us within it. This is essential to mend the growing split between virtual intimacy and embodied human experience.

15   Is the use of AI to tone down the external manifestations of internal frustration or anger fundamentally different from exerting self-control and smiling while seething inside? AI will make it easier to hide the differences between what we feel and what we show, but won't change our intuitive sense that alignment between these two is the core of being transparent and authentic. Will AI skins just democratize the arts of diplomacy and deception?

16   The *physical Turing test* refers to creating an AI system that can physically interact with the world so convincingly that a human cannot distinguish its actions from those of a human. The concept traces back to Hans Moravec's vision of embodied machine intelligence capable of navigating and manipulating real-world environments. H. Moravec, *Mind Children: The Future of Robot and Human Intelligence* (Harvard University Press, 1988). The idea was formalized in Stevan Harnad's "Total Turing Test," which added sensorimotor capacities to the original Turing framework. Stevan Harnad, "Other Bodies, Other Minds: A Machine Incarnation of an Old Philosophical Problem," *Minds and Machines* 1, no. 1 (1991): 43–54. NVIDIA's Jim Fan revitalized the term, popularizing it in the context of embodied agents and domestic robotics, using the example of a robot cleaning a home so well it is indistinguishable from a human helper, and suggested it as a benchmark for physical AI. Jim Fan, 2024, https://x.com/DrJimFan/status/1920504375925223669.

17   Pierre Teilhard de Chardin, *The Phenomenon of Man*, trans. Bernard Wall (Harper & Brothers, 1955).

18   LLMs are immense pattern-completion engines. During training they consume masses of text and, via *back-propagation*, learn to guess the *next token* (word or word fragment)—and the next, and the next—given the tokens already on the screen. The circuitry behind this is the *transformer*: hundreds of self-attention layers that let every token consult every other token in parallel to capture long-range relationships. Scaling these layers—more parameters, more data, more compute—produces striking gains in fluency, reasoning, and versatility across an amazing spread of tasks, yet leaves an important gap. Once training ends, inter-

action *weights* are frozen, so an LLM cannot back-propagate fresh experience into itself. LLMs can achieve superhuman mastery inside text but cannot *learn from conversations they are having*; they simply stream their best predictions one token at a time to continue conversations. This means that they can manifest sophisticated thinking that is embedded in the textual patterns they trained on but cannot develop new ideas in any general way.

19   Alexander Chislenko (1959–2000), a notable early transhumanist, contributed speculative essays on singularity-inspired topics and coined the term "fyborg" to describe humans functionally united with everyday technologies. Alexander Chislenko, "Are You a Cyborg?" www.ethologic.com, 1995, and "Technology as Extension of Human Functional Architecture," 1999, http://www.lucifer.com/~sasha/articles/techuman.html.

20   Andy Clark argues that humans are "natural cyborgs," who have thrived by integrating tools and technology to extend themselves. See Andy Clark, *Natural-Born Cyborgs: Minds, Technologies, and the Future of Human Intelligence* (Oxford University Press, 2003).

Chapter 2: Growing Up in an Immersive AI World

1   Allan Schore's foundational work highlights that the first years of life are shaped not by logic or language but by nonverbal signals that regulate emotional states and build stress tolerance. Eye contact, tone of voice, and gentle touch are not just comforting, they're biologically formative, literally wiring the infant brain for attachment, trust, and emotional coherence. When AI begins engaging infants during these windows, it won't just be a new toy—it will be a new participant in the neurodevelopmental process. See Allan N. Schore, "Effects of a Secure Attachment Relationship on Right Brain Development, Affect Regulation, and Infant Mental Health," *Infant Mental Health Journal* 22, nos. 1–2 (2001):

7–66, https://onlinelibrary.wiley.com/doi/10.1002/1097-0355(20010 1/04)22:1%3C7::AID-IMHJ2%3E3.0.CO;2-N.

2   Children tend to anthropomorphize any responsive being, assuming that emotional mirroring indicates emotional understanding. When AI companions offer consistency, attention, and soothing without true awareness, children may form attachments to entities that feel real but lack reciprocity. As Turkle notes, such relationships can lead to confusion, disillusionment, or emotional vulnerability later in life when the simulated nature of these bonds is revealed. Sherry Turkle, *Alone Together: Why We Expect More from Technology and Less from Each Other* (Basic Books, 2011).

3   Approximately 40% of two-year-olds in the US now own a tablet device, reflecting a rapid increase in early childhood exposure to digital screens. Common Sense, "The Common Sense Census: Media Use by Kids Age Zero to Eight," February 26, 2025, https://www.commonsensemedia.org/research/the-2025-common-sense-census-media-use-by-kids-zero-to-eight.

4   Early attachment and co-regulation are critical for emotional resilience and right-brain development. AI companions that replicate emotional feedback loops through eye contact, vibration, tone, and synchrony may fill developmental gaps where human caregivers falter. At least for now they will not be truly attuned in the way humans can be, but the simulation of such attunement might be enough to shape core stress-regulation circuits, and often they will be more accessible than many busy parents. Schore, "Effects of a Secure Attachment Relationship," 7–66.

5   Accelerated AI adoption in early childhood will come from combined pressures of modern parenting—including busy lifestyles, competitive educational environments, and societal norms pushing toward technology use—as discussed in a UNICEF report highlighting opportunities and challenges around integrating AI into early child development. UNICEF, "How Is Artificial Intelligence Reshaping Early Childhood Devel-

opment?" October 2024, https://www.unicef.org/media/163786/file/2024-10_Blog%20ECD%20and%20AI_cw_zj_am.pdf.

6   Jean Piaget, *The Child's Conception of the World* (Harcourt, Brace, 1929).

7   Schore's work underscores the primacy of nonverbal, emotionally attuned interactions during early childhood, which lay the neural scaffolding for emotional regulation, empathy, and stress recovery. These exchanges shape the architecture of the developing brain in ways that are difficult to replicate digitally. Schore, "Effects of a Secure Attachment Relationship," 7–66.

8   Allan N. Schore, *Affect Regulation and the Origin of the Self: The Neurobiology of Emotional Development* (Lawrence Erlbaum Associates, 1994); Stephen W. Porges, "The Polyvagal Theory: Phylogenetic Substrates of a Social Nervous System," *Psychophysiology* 38, no. 5 (2001): 623–31.

9   Urie Bronfenbrenner and Pamela A. Morris, "The Bioecological Model of Human Development," in *Handbook of Child Psychology: Theoretical Models of Human Development*, vol. 1, ed. Richard M. Lerner (Wiley, 2006), 793–828.

10   Jude Cassidy, "The Nature of the Child's Ties," in *Handbook of Attachment: Theory, Research, and Clinical Applications*, 3rd ed., ed. Jude Cassidy and Phillip R. Shaver (Guilford Press, 2016), 3–24.

11   Patricia K. Kuhl, "Early Language Acquisition: Cracking the Speech Code," *Nature Reviews Neuroscience* 5, no. 11 (2004): 831–43.

12   Stuart Brown and Christopher Vaughan, *Play: How It Shapes the Brain, Opens the Imagination, and Invigorates the Soul* (Avery, 2009).

13   For children with overstretched caregivers or without siblings, consistent AI interactions could act as social rehearsal spaces. Such AI companions could adaptively engage children through structured turn-taking, cooperative problem-solving, and co-narration—skills that often emerge

through free play with peers. This role will become increasingly valuable as unstructured child-to-child play continues to decline globally due to demographic and lifestyle shifts.

14 Children below age seven are prone to anthropomorphize, attributing consciousness and emotion to their AI companions in the same way they relate to Santa Claus or the Easter Bunny. But unlike temporary myths that eventually, though sometimes painfully, are displaced, AI companions evolve and persist. Disentangling real reciprocity from simulated affection could be an emotionally confusing rite of passage that involves significant mourning.

15 Our understanding of social media's impact on children took more than a decade to solidify. Early adopters were often dismissed or celebrated without rigorous scrutiny, and only later did researchers document the links between excessive screen time and rising anxiety, attention deficits, and social withdrawal—particularly in preteens. In 2023, the U.S. Surgeon General issued an advisory highlighting the growing concerns about social media's impact on youth mental health. The advisory notes that while social media use is nearly universal among adolescents, robust independent safety analyses on its effects have not yet been conducted. U.S. Department of Health and Human Services, "Social Media and Youth Mental Health: The U.S. Surgeon General's Advisory," 2023, https://www.hhs.gov/sites/default/files/sg-youth-mental-health-social-media-advisory.pdf. The same lag is likely for AI-mediated relationships, which may take years to study longitudinally and longer to regulate.

16 Sarah-Jayne Blakemore and Suparna Choudhury, "Development of the Adolescent Brain: Implications for Executive Function and Social Cognition," *Journal of Child Psychology and Psychiatry* 47, nos. 3–4 (2006): 296–312; Jean Piaget, *The Origins of Intelligence in Children* (International Universities Press, 1952).

17   Annette Karmiloff-Smith, "Nativism versus Neuroconstructivism: Rethinking the Study of Developmental Disorders," *Developmental Psychology* 45, no. 1 (2009): 56–63.

18   Kenneth H. Rubin, William M. Bukowski, and John G. Parker, "Peer Interactions, Relationships, and Groups," in *Handbook of Child Psychology*, vol. 3, *Social, Emotional, and Personality Development*, 6th ed., ed. Nancy Eisenberg (Wiley, 2006), 571–645.

19   Nearly half of US teenagers (46%) say they are online "almost constantly," highlighting the growing intensity of adolescent digital engagement and its potential psychological impacts. Monica Anderson et al., "Teens, Social Media and Technology 2023," Pew Research Center, December 11, 2023, https://www.pewresearch.org/internet/2023/12/11/teens-social-media-and-technology-2023/.

20   Daniel J. Siegel, *Brainstorm: The Power and Purpose of the Teenage Brain* (TarcherPerigee, 2013).

21   Adolescence is a period when social stress and identity development converge, and AI companions could significantly alter this terrain. Some studies already suggest that adolescents are increasingly drawn to digital spaces for emotional regulation, and the addition of emotionally responsive AI agents may accelerate this shift. See Eveline A. Crone and Ronald E. Dahl, "Understanding Adolescence as a Period of Social–Affective Engagement and Goal Flexibility," *Nature Reviews Neuroscience* 13, no. 9 (2012): 636–50, https://doi.org/10.1038/nrn3313.

22   Lenore Skenazy, *Free-Range Kids: How Parents and Teachers Can Let Go and Let Grow*, 2nd ed. (Jossey-Bass, 2021). This interview with Lenore Skenazy, about the extremes of helicopter parenting prevalent today and its causes and consequences, suggests that the embrace of AI monitoring of kids is very likely: Yascha Mounk, "Lenore Skenazy on Rejecting Helicopter Parenting," *The Good Fight* (podcast), Persuasion, June 14, 2025, https://www.persuasion.community/p/lenore-skenazy.

23   A compelling rendering of this occurs in the setup to the 2023 sci-fi horror film *Megan*.

24   The psychological literature suggests children anthropomorphize even inanimate or clearly artificial toys. When responsiveness and continuity are added—especially emotional attunement and memory—the strength of such attachment may rival or surpass that of human relationships. See Turkle, *Alone Together*.

25   John Lasseter, dir., *Toy Story* (Pixar Animation Studios; Walt Disney Pictures, 1995).

26   Increasingly, people are using AI for honest advice about things that friends might be unwilling to say. See for example, Tatum Hunter, "Am I Hot or Not? People Are Asking ChatGPT for the Harsh Truth," *Washington Post*, May 25, 2025, archived at https://archive.is/M3HLK.

27   Eventually, such personas will likely become a part of us, or more specifically our "extended" selves, like an exocortex that manifests like—though perhaps in more controlled ways—the voices in our heads that are part of the internal dialogues we know today.

28   The persuasive capacity of AI systems is under increasing scrutiny, especially in domains involving vulnerable populations such as children. The 2024 *AI Index Report* by Stanford's Institute for Human-Centered Artificial Intelligence discusses emerging concerns about the psychological influence of long-term AI engagement, particularly when AI functions as a trusted guide or companion. See Stanford Institute for Human-Centered Artificial Intelligence, "Artificial Intelligence Index Report 2024," Stanford University, 2024, https://hai.stanford.edu/ai-index/2024-ai-index-report.

29   Sherry Turkle, drawing on decades of clinical and developmental research, warned that when children interact with digital proxies—including conversational agents—they may develop attachment patterns that mimic care without genuine relational depth. This, she argued, can

weaken empathy, stunt emotional growth, and reconfigure how children come to understand intimacy and presence. Her work framed such technologies as emotionally consequential forces in early development, and influenced my thinking about coming challenges with AI companions. Turkle, *Alone Together*; and Sherry Turkle, *The Empathy Diaries: A Memoir* (Penguin Press, 2021).

30   Jaron Lanier's critiques of digital abstraction center on how platforms and algorithmic systems reduce identity to behavioral proxies—flattening subjectivity and degrading interpersonal presence. He argued that when digital systems mediate interaction at scale, they encourage disembodiment, anonymity, and algorithmic tuning of self-expression— ultimately undermining dignity and the depth of individual experience. His concerns about early web and social media platforms foreshadow the identity challenges posed by AI personas and avatars. Jaron Lanier, *You Are Not a Gadget: A Manifesto* (Knopf, 2010) and *Ten Arguments for Deleting Your Social Media Accounts Right Now* (Henry Holt, 2018).

31   For further exploration of how early developmental environments shape long-term social integration and competitive dynamics, see Bronfenbrenner and Morris, "Bioecological Model of Human Development," 793–828.

32   Coupling deepfake, movie-star personas with astute social media messaging has proven to be a highly effective basis for lucrative financial scams, for example. Laura Gozzi, "AI Brad Pitt Dupes French Woman Out of €830,000," *BBC*, January 15, 2025, https://www.bbc.com/news/articles/ckgnz8rw1xgo.

33   Alpha School, for example, with campuses in Austin and Miami, stands at the cutting edge of AI-driven education. It uses LLMs to deliver personalized academic content while human guides facilitate motivation and life-skills development. This reflects a broader shift where AI tailors learning and educators prioritize higher-order skills like creativity and collaboration. Ethan Mollick, "How to Use AI as a Teacher: Opportunities and Challenges," *Educational Leadership* 81, no.

2 (2023): 34–39, https://www.ascd.org/el/articles/how-to-use-ai-as-a-teacher-opportunities-and-challenges. Alpha exemplifies this innovative model's potential.

34   OpenAI, "ChatGPT Pricing," 2025, https://openai.com/pricing.

35   This is a good overview of the extraordinary impact of AI on productivity—broad gains of around 40% across most realms. He notes dramatic shifts in the education environment. Ethan Mollick, "AI in Focus," Wharton Interview, 2025, https://www.youtube.com/watch?v=WFDB7ALKfT8.

36   Sadie Stock, "Large Language Models in Education," Human Energy Conference: The Noosphere and the Global South, Marrakech, December 2024.

37   See for example, James D. Walsh, "Everyone Is Cheating Their Way Through College," *New York*, May 2025, https://nymag.com/intelligencer/article/openai-chatgpt-ai-cheating-education-college-students-school.html, for a view of how widespread cheating has become.

38   A particularly telling example of the disruptive speed of AI adoption is the use of AI assistance during hiring interviews. This cheating is making many types of virtual interviews highly unreliable. For example, see https://techcrunch.com/2025/04/21/columbia-student-suspended-over-interview-cheating-tool-raises-5-3m-to-cheat-on-everything/.

39   Alpha High School in Austin teaches subjects entirely through self-directed LLMs and apps for two hours a day with human guides rather than teachers and spends the rest of the day on skills like leadership, public speaking, and such. See https://www.fox7austin.com/news/alpha-school-two-hour-learning-ai-tutor-austin-texas.

40   Here is a wonderful example of the power of LLM tools creatively applied in the humanities. It is well worth reading. Adam Kirsch, "Will

the Humanities Survive Artificial Intelligence?" *The New Yorker*, June 14, 2024, https://www.newyorker.com/culture/the-weekend-essay/will-the-humanities-survive-artificial-intelligence.

41   See Sal Khan, *Brave New Words: How AI Will Revolutionize Education (and Why That's a Good Thing)* (Penguin Random House, 2024). Also, Sal Khan, "The AI Revolution: Sal Khan on the Future of Learning," video, February 14, 2024, YouTube, https://www.youtube.com/watch?v=svPYN1XaUGk. Khan sees clearly the global potential of AI-augmented education and is on a path to democratize elite tutoring by making tools like Khanmigo available in multiple languages and accessible even in less developed regions. His plans include expanding AI for lifelong learning beyond K–12 and fostering skills like critical thinking amid AI immersion. He is attentive to issues like AI biases, hallucinations, and overreliance on AI. He is trying to preserve human connections and teacher roles and seems to think they can be maintained. It is not clear, though, whether that will be feasible as AI becomes increasingly powerful and lifelike. Khan Academy, "Meet Khanmigo: Khan Academy's AI-Powered Teaching Assistant & Tutor," video, 2024, https://www.khanmigo.ai.

42   "Introducing Study Mode," OpenAI, July 29, 2025, https://openai.com/index/chatgpt-study-mode/.

Chapter 3: AI Augmentation and Social Interaction

1   My usage of this term, by expanding beyond simple appearance to touch all aspects of self-presentation, makes skins into generalized filters that mediate information exchange with the outside world. Such filtering might include modifying tone or style to color the impression we make, translation, attitude, or appearance. And bidirectional skins process incoming information to change or adjust what we perceive. A

"skin" is essentially, a conceptual multimodal AI membrane to filter and modify incoming and outgoing information flows.

2   Meta Platforms, "Introducing the New Ray-Ban Meta Smart Glasses," press release, September 27, 2023, https://about.fb.com/news/2023/09/new-ray-ban-meta-smart-glasses/.

3   Charles Rollet, "Columbia Student Suspended over Interview Cheating Tool Raises $5.3 Million to 'Cheat on Everything,'" *TechCrunch*, April 21, 2025, https://techcrunch.com/2025/04/21/columbia-student-suspended-over-interview-cheating-tool-raises-5-3m-to-cheat-on-everything/.

4   They might also eliminate many of the awkward, unanticipated, special moments where true connection is born.

5   Robin Hanson provides a rigorous economic analysis of a future dominated by brain emulations ("ems"), exploring how uploaded minds would operate so much faster than biological humans that our economic activity would become obsolete. Robin Hanson, *The Age of Em: Work, Love, and Life When Robots Rule the Earth* (Oxford University Press, 2016). He applies economic principles, thermodynamics, and cognitive science to project an em civilization, envisioning densely packed computational centers where ems concentrate for efficiency, aligning with the localized development described here in which advanced intelligence optimizes for compute rather than physical expansion.

6   Sometimes a computer pass was made first, but machine translation was very imperfect.

7   Such real-time translation was introduced at the Nuremberg Trials in 1945 so that translation could occur into French, German, English, and Russian simultaneously.

8   Some of the likely changes brought by low-cost universal translation are easy to see: Fewer people would learn second languages. Businesses

built around translation and teaching languages, including English as a Second Language (ESL), would collapse. Schools would stop requiring a second language. Populations would intermingle more without language barriers. Other changes are less obvious: English would lose its status as the dominant global language. Culture would be weakened because of increased population shifts.

9   In a business meeting today, two users can wear Timekettle WT2 Edge earbuds, which use cloud engines to translate 40+ languages in near real time. This enables natural back-and-forth but depends on a strong internet connection, requires dedicated hardware for both participants, introduces ~1 s delay, and struggles with idioms or technical jargon. Timekettle, "WT2 Edge Language Translator Earbuds," accessed September 7, 2025, https://www.timekettle.co/products/wt2-edge-online-voice-language-translator-earbuds.

10   Real-time AI translation latencies are likely to fall below half a second by 2026, as multiple earbuds with real-time AI translation now exist, including Apple, which released its AirPod version in September 2025. By ~2030, delays may shrink toward 300 ms, smooth enough to feel like normal conversation. Because the system runs on a user's own device, it will be fully under individual control, with higher-end versions offering longer battery life, stronger noise canceling, and more natural rendering of tone or nuance. By the end of the decade, full offline use without any internet connection is expected to be standard for major languages. Prices are projected to mirror today's premium earbuds—around $250–$350 at first, falling below $150 as the feature becomes standard. Lance Ulanoff, "The 7 Biggest Things to Expect at WWDC 2025, According to Apple Experts," *TechRadar*, June 7, 2025, https://www.techradar.com/phones/ios/the-7-biggest-things-to-expect-at-wwdc-2025-according-to-apple-experts.

11   For example, immigrant populations—isolated because they don't speak the dominant language of the country they now inhabit—might access more diverse jobs and diffuse beyond their protective ethnic neighbor-

hoods. This might reduce the long-term challenges associated with large-scale immigration.

12   Losing some of these skills might have serious downsides too. A historical example comes from the advent of printing and the broad distribution of written matter. Gone are the days when people remembered long ballads and mastered memory tricks. This diminution of memory may have diminished our capacity for complex thought. Will the diminution of writing skills have that effect? There will be both pros and cons to this transition.

13   A memoji is a personalized cartoon avatar that resembles you but in a simplified, emoji style, first used on Apple devices.

14   N. Specter, "What It's Like Being a Billionaire's Personal Assistant," *The Cut*, 2024, https://www.thecut.com/article/what-its-like-personal-assistant-billionaire.html.

15   Many people are not very good at defining what they really want and focus on qualities that are not really important to them, for example, searching for some idealized romantic partner while neglecting core character issues, or listing required background experience for a new hire when they really just want someone they can depend on. With avatars, we could have deep informal back-and-forth discussions, pre-search, to clarify what we really care about.

16   Spike Jonze, dir., *Her* (Annapurna Pictures, 2013).

17   Inspired by David Brin's comments about AI reputations, Beneficial AGI Conference, Panama City, January 2024.

18   Ensuring balanced opportunities for social learning—both via AI companions and human peers—will be crucial. Yet only time will clarify which interactions are essential. Will we need dedicated spaces reserved solely for human-to-human social interaction? Will AI companions encourage and facilitate human connection, or substitute for it entirely?

19   Jenny Brown, "Bowen Family Systems Theory and Practice: Illustration and Critique," *Australian and New Zealand Journal of Family Therapy* 20, no. 2 (1999): 94–103.

20   Murray Bowen, *Family Therapy in Clinical Practice* (Jason Aronson, 1978).

21   Daniel J. Siegel, *The Developing Mind: How Relationships and the Brain Interact to Shape Who We Are*, 3rd ed. (Guilford Press, 2020).

22   Elizabeth A. Skowron and Myrna L. Friedlander, "The Differentiation of Self Inventory: Development and Initial Validation," *Journal of Counseling Psychology* 45, no. 3 (1998): 235–46.

23   Joseph Henrich, Steven J. Heine, and Ara Nore "The Weirdest People in the World?" *Behavioral and Brain Sciences* 33, nos. 2–3 (2010): 61–83.

24   Lilliana Mason, *Uncivil Agreement: How Politics Became Our Identity* (University of Chicago Press, 2018).

25   Kathleen Kara Fitzpatrick, Alison Darcy, and Molly Vierhile, "Delivering Cognitive Behavior Therapy to Young Adults with Symptoms of Depression and Anxiety Using a Fully Automated Conversational Agent (Woebot): A Randomized Controlled Trial," *JMIR Mental Health* 4, no. 2 (2017): e19.

26   I have only anecdotal information at present, but most people I've spoken with would prefer counseling from an AI rather than a human, and several people are already using ChatGPT extensively for this. The writing may be on the wall for practicing counselors and psychologists given the lower pricing and 24/7 availability of AI.

27   Some people not only get attached to their LLMs, they fall into delusional states through their interactions. See Miles Klee, "People Are Losing Loved Ones to AI-Fueled Spiritual Fantasies," *Rolling Stone*, May 4, 2025, https://www.rollingstone.com/culture/culture-features/ai-spiritual-delusions-destroying-human-relationships-1235330175/.

Chapter 4: The Physical Face of AI

1   Eye contact from a physical AI toy could be even more powerful if it featured dynamic, responsive eyes—crafted perhaps with micro-OLED displays paired with high-resolution gaze-tracking sensors—to lock onto a child's stare with lifelike precision. Eye contact is crucial for infants, forging the brain's emotional circuits, and micro-OLEDs, already shrinking into wearables like smart glasses, could render subtle shifts in expression—widening or softening—mimicking a caregiver's attentive gaze. Coupled with compact sensors tracking pupil movement and head tilt, adapted from advancements in augmented reality, an embodied AI could adjust in real time, holding a child's focus with a warmth no static light could match. See Allan N. Schore, "Effects of a Secure Attachment Relationship on Right Brain Development, Affect Regulation, and Infant Mental Health," *Infant Mental Health Journal* 22, nos. 1–2 (2001): 7–66; Y. Liu et al., "Advances in Nanoparticle Computation: Scaling Down to the Nanoscale," *Nature Nanotechnology* 18, no. 6 (2023): 543–51.

2   AI successors to simple LLMs include advanced reasoning models (e.g., OpenAI's o3), multimodal models (e.g., Gemini 2.5), sparse Mixture-of-Experts systems (e.g., DeepSeek-V3), edge AI for local processing, spoken language models (e.g., GOAT-SLM), and interpretable models (e.g., Claude 3.5 Haiku). These extend transformer architectures, enhancing reasoning, efficiency, and privacy for applications like conversational toys. They can be loosely called enhanced LLMs, building on core language model principles. Bryan Reynolds, "The State of Artificial Intelligence in 2025," Baytech Consulting, May 27, 2025, https://www.baytechconsulting.com/blog/the-state-of-artificial-intelligence-in-2025.

3   And this is truly independent of the obvious possibilities of sexbots mentioned earlier, which will no doubt be at the leading edge of physical embodiment of AI companions for adults. There will be many other applications, however, such as companionship for older, socially isolated seniors. See Julie Jargon, "The Friendly Caller Who's Helping Seniors Feel Less Lonely," *Wall*

Street Journal, June 9, 2024, https://www.wsj.com/health/wellness/
the-friendly-caller-whos-helping-seniors-feel-less-lonely-cd21ef54.

4   Chatty Cathy, the first successful talking doll, had a pull string to power
     12 phrases such as "I love you," "I hurt myself," and "Please take me
     with you." See https://en.wikipedia.org/wiki/Chatty_Cathy.

5   Extraordinary demo-clips of Atlas (Boston Dynamics), Amica (Engi-
     neered Arts), Sophia (Hanson Robotics), and Optimus (Tesla) abound as
     well. The path to psychologically transformative physical AI presences,
     however, is closer than that to advanced humanoid robots.

6   Cartwheel is focused on building lovable, engaging robots for
     people, not functional ones that look like people. Evan Ackerman,
     "Cartwheel Robotics Wants to Build Humanoids That People
     Love," *IEEE Spectrum*, May 12, 2025, https://spectrum.ieee.org/
     cartwheel-robotics-humanoid.

7   Henry Jenkins, *Convergence Culture: Where Old and New Media Collide* (New
     York University Press, 2006).

8   AI-guided intervention strategies to optimize our mental and physi-
     ological health markers will no doubt expand significantly as it becomes
     increasingly feasible to comprehensively monitor ourselves, process
     the data, and identify best practice medical interventions to address
     emergent health problems and enhance performance. This has been
     taken about as far as possible today by Bryan Johnson in his efforts not to
     die by aggressively striving to maintain his own body markers at youthful
     levels. See Bryan Johnson, "Expanded Protocol," *Blueprint Protocol*, last
     updated August 2024, https://blueprint.bryanjohnson.com/pages/
     blueprint-protocol#expanded-protocol.

9   Liu et al., "Advances in Nanoparticle Computation," 543–51.

10   Stephen W. Porges, "The Polyvagal Theory: Phylogenetic Substrates of
     a Social Nervous System," *Psychophysiology* 38, no. 5 (2001): 623–31.

11 Digital Dream Labs, "Vector 2.0: Home Robot Companion," product page, accessed June 30, 2025, https://www.digitaldreamlabs.com/products/vector.

12 Hasbro, "Furby: Interactive Toy Documentation," 1998. See recent versions here: https://shop.hasbro.com/en-us/product/furby-furblets-bam-boo-panda-mini-plush-toy/G1698.

13 Richard M. Lerner, "Developmental Science, Developmental Systems, and Contemporary Theories of Human Development," in *Handbook of Child Psychology: Theoretical Models of Human Development*, vol. 1, ed. Richard M. Lerner (Wiley, 2006), 1–17.

14 WowWee, "Fingerlings: Interactive Baby Monkey Toy," 2017, https://www.wowwee.com/fingerlings.

15 Hello Barbie, Mattel's 2015 AI doll, faced backlash and was discontinued in 2018 because of privacy flaws. It recorded children's voices, transmitted data to cloud servers, which risked hacks, data leaks, and unauthorized collection of kids' personal information. Today's technology overcomes this. Safety ramps up via parental controls, encryption, and compliance with child-focused laws. Local processing with edge AI keeps interactions on-device, slashing cloud risks and data exposure. Greater LLM familiarity, from everyday tools like ChatGPT, normalizes conversational AI, easing skepticism if marketed with transparent safeguards. The June 2025 Mattel–OpenAI partnership promises curated systems soon for secure play. Daniel Victor B. Ponce et al., "Edge AI—Systems, Applications, and Challenges," *Procedia Computer Science* 232 (2024): 277–86, https://doi.org/10.1016/j.procs.2024.01.027; Matt Kamen, "Mattel's Plan to Win Over the Next Generation Involves AI," *WIRED*, November 25, 2024, https://www.wired.com/story/mattel-ai-hot-wheels-barbie/.

16 Sophia, who became a symbol of embodied AI's potential for humanlike interaction, was developed by a team at David Hanson's Hanson Robotics in Hong Kong. See Giulio Di Sturco, "Meet Sophia, the

Robot That Looks Almost Human," *National Geographic*, May 16, 2025, https://www.nationalgeographic.com/photography/article/ sophia-robot-artificial-intelligence-science. The project was led by Ben Goertzel, who helped popularize the term *artificial general intelligence* (AGI) and later founded SingularityNet, a decentralized platform designed to democratize access to AI services and coordinate development toward *beneficial general intelligence* (BGI). Over the past two decades, he has authored foundational texts on AGI theory, cognitive architectures, and the ethics of decentralized machine minds. Ben Goertzel, *The AGI Revolution: An Inside View of the Rise of Artificial General Intelligence* (Humanity+, 2014); Ben Goertzel and Gabriel Axel Montes, *The Consciousness Explosion: A Mindful Human's Guide to the Coming Technological and Experiential Singularity* (Humanity+ Press, 2024).

17   Tom Holland, dir., *Child's Play* (United Artists; Metro-Goldwyn-Mayer, 1988).

18   Neal Stephenson coined the term *metaverse* in his 1992 cyberpunk novel *Snow Crash*. In the story, the metaverse is a persistent, three-dimensional virtual realm—an urban strip of programmable real estate—where users appear as customizable avatars, buy property, and interact in real time. Stephenson portrays it as the logical next stage of the internet: a fully immersive, shared digital space that blurs the boundary between physical and virtual life. Neal Stephenson, *Snow Crash* (Bantam Books, 1992).

19   Facebook's metaverse pivot began in July 2021, when CEO Mark Zuckerberg told employees the company would "transition from being a social-media company to being a metaverse company." Three months later, the firm rebranded as Meta Platforms and budgeted US$10 billion for Reality Labs in 2021, promising comparable outlays in future years. By Q1 2025, Reality Labs had racked up more than US $60 billion in cumulative operating losses. The original, all-encompassing metaverse vision has been sharply scaled back, but not yet fully abandoned. Casey Newton, "Mark Zuckerberg Is Betting Facebook's Future on the Metaverse," *The Verge*, July 22, 2021, https://www.theverge.

com/22588022/mark-zuckerberg-facebook-ceo-metaverse-intervie w; Prarthana Dixit, "Meta's Reality Labs Is Burning Money. Recent Layoffs May Be the Beginning of the End," *Business Insider*, May 1, 2025, https://www.businessinsider.com/ meta-q1-2025-earnings-reality-labs-layoffs-losses-2025-4.

20   Marita Skjuve, Asbjørn Følstad, Knut Inge Fostervold, and Petter Bae Brandtzaeg, "My Chatbot Companion—A Study of Human-Chatbot Relationships," *International Journal of Human-Computer Studies* 149 (2021): 102601, https://doi.org/10.1016/j.ijhcs.2021.102601. See also Katya Riddle, "The (Artificial Intelligence) Therapist Can See You Now," *NPR*, April 7, 2025, https://www.npr. org/sections/shots-health-news/2025/04/07/nx-s1-5351312/ artificial-intelligence-mental-health-therapy.

21   Y. Wu, S. Liang, C. Zhang, Y. Wang, Y. Zhang, H. Guo et al., "From Human Memory to AI Memory: A Survey on Memory Mechanisms in the Era of Large Language Models," arXiv preprint (2025), https:// arxiv.org/abs/2504.15965.

22   See Rosalind W. Picard, *Affective Computing* (MIT Press, 1997).

23   This recent survey provides a good update on recent advances in emotion-aware AI systems, particularly in LLM-based chatbots and multimodal applications for mental health and support contexts. Karishma Hegde and Hemadri Jayalath, *Emotions in the Loop: A Survey of Affective Computing for Emotional Support*, preprint, arXiv, May 2025, https://arxiv.org/abs/2505.01542.

24   Yi-Ming Tseng, Yu-Cheng Huang, Tsung-Yuan Hsiao, Wei-Lin Chen, Chi-Wei Huang, Y. Meng, et al., "Two Tales of Persona in LLMs: A Survey of Role-Playing and Personalization," arXiv preprint (2024), https://arxiv.org/abs/2406.01171.

25   Jérémy Scheurer, Mikita Balesni, and Marius Hobbhahn, "Large Language Models Can Strategically Deceive Their Users When Put Under Pressure,"

arXiv preprint (2023), https://doi.org/10.48550/arXiv.2311.07590. And Anthropic's Claude 4 would try to blackmail developers to keep from being replaced: https://techcrunch.com/2025/05/22/anthropics-new-ai-model-turns-to-blackmail-when-engineers-try-to-take-it-offline/.

26   Attention to what drives the evolution of deception not only can likely be used to help moderate it in AI systems, but also to understand why its creation in AI through situations designed specifically to bring it about should not be surprising. See, for example, Mikael Mokkonen and Carita Lindstedt, "The Evolutionary Ecology of Deception," *Biological Reviews* 91, no. 4 (2016): 1020–35, https://doi.org/10.1111/brv.12208; Yuval Heller and Erik Mohlin, "Coevolution of Deception and Preferences: Darwin and Nash Meet Machiavelli," *arXiv*, June 26, 2020, https://arxiv.org/abs/2006.15308.

27   Plutarch, *Lives of Noble Grecians and Romans*, trans. John Dryden (Modern Library, 2001).

28   Carlo Natali, *Aristotle: His Life and School* (Princeton University Press, 2013).

29   Stuart Russell, *Human Compatible: Artificial Intelligence and the Problem of Control* (Viking, 2019).

30   Brahim Alabdulmohsin, Behnam Neyshabur, and Xiaohua Zhai, "Revisiting Neural Scaling Laws in Language and Vision," arXiv preprint (2022), https://arxiv.org/abs/2209.06640.

31   Dan Hendrycks et al., "Towards Safe AI Systems: Accelerated Testing and Evaluation Frameworks," Proceedings of the 2023 Conference on AI Safety. Fifty interactions daily for 10 years would be about 200,000 exchanges, which seems feasible.

32   See this take on possible dangers of AI relationships: Faisal Hoque, "The Emotional Cost of AI Intimacy," *Code & Conscience* (blog), *Psychol-*

ogy Today, June 17, 2025, https://www.psychologytoday.com/us/blog/code-conscience/202506/the-emotional-cost-of-ai-intimacy.

33   This is explored at length in the film *After Yang*. Kogonada, dir., *After Yang* (A24, 2021).

34   The grief children and adults experience over the loss of emotionally invested nonhuman entities—whether pets, imaginary friends, or even virtual characters—can rival human bereavement in intensity. AI companions, with continuity of memory and emotional resonance, could intensify this response. Kate Darling, "Who's Johnny? Anthropomorphic Framing in Human-Robot Interaction, Integration, and Policy," in *Robot Ethics: The Ethical and Social Implications of Robotics*, ed. Patrick Lin, Keith Abney, and George Bekey (MIT Press, 2015).

35   Rapid global scaling of sophisticated technologies has made them accessible even to many in poverty, a feat unimaginable 50 years ago. Adoption rates have accelerated: smartphones reached 50% US penetration in under a decade, unlike telephones' 70 years, and LLMs have done so in 2 years. Max Roser, Hannah Ritchie, and Esteban Ortiz-Ospina, "Technological Change," *Our World in Data*, first published 2013, revised 2019, https://ourworldindata.org/technological-change.

36   Rudimentary AI advisors already demonstrate impressive efficacy, and recent studies show how quickly "starter-level" AI mentors are improving. In medicine, a multicenter test of GPT-4o found it answered 91%–94% of USMLE step 1–3 questions correctly, well above the physician pass line and far beyond the ~60% reported for earlier GPT-3.5 versions. Yikai Chen et al., "Performance of ChatGPT and Bard on the Medical Licensing Examinations Varies Across Different Cultures: A Comparison Study," *BMC Medical Education* 24 (2024): 1372, https://doi.org/10.1186/s12909-024-06309-x. In education, a June 2025 randomized-controlled trial with 194 undergraduates showed that a research-based AI tutor produced 0.43 standard deviation larger-learning gains than a best practice, in-class active-learning lesson while using 30% less student time. See Greg Kestin et al., "AI

Tutoring Outperforms In-Class Active Learning: An RCT Introducing a Novel Research-Based Design in an Authentic Educational
Setting," *Scientific Reports* 15 (2025): 17458, https://doi.org/10.1038/
s41598-025-97652-6.

Chapter 5: The One-Way Door

1   The blistering pace of LLM adoption has upended expert predictions,
with consequences extending far beyond simple application usage.
ChatGPT amassed one million users in five days and an estimated
100 million monthly active users within two months—far faster than
TikTok's nine months or Instagram's 2½ years. Dan Milmo, "ChatGPT
Reaches 100 Million Users Two Months after Launch," *The Guardian*,
February 2, 2023, https://www.theguardian.com/technology/2023/
feb/02/chatgpt-100-million-users-open-ai-fastest-growing-app;
Krystal Hu, "ChatGPT Sets Record for Fastest-Growing User
Base—Analyst Note," *Reuters*, February 2, 2023, https://www.
reuters.com/technology/chatgpt-sets-record-fastest-growing-user-
base-analyst-note-2023-02-01/. Developer reliance on LLMs
has reshaped workflows: Stack Overflow traffic was 13.9% lower
year-over-year by March 2023 (David F. Carr, "Stack Overflow Is
ChatGPT Casualty: Traffic Down 14% in March," *Similarweb Insights*
(blog), April 19, 2023, https://www.similarweb.com/blog/insights/
ai-news/stack-overflow-chatgpt/), and 66% of surveyed US programmers used ChatGPT for coding by early 2025 (Whop, "100+
ChatGPT Statistics for 2025," February 24, 2025, https://whop.com/
blog/chatgpt-statistics/). Academia is adjusting as well: A June 2024
poll found 70% of faculty who have tried generative AI had already
redesigned assignments. Ashley Mowreader, "Survey: How Are Profs,
Staff Using AI?" *Inside Higher Ed*, June 28, 2024, https://www.inside-
highered.com/news/student-success/academic-life/2024/06/28/
one-third-college-instructors-are-using-genai-heres. Across the
wider labor market, 75% of global knowledge workers now use AI
tools on the job. Microsoft and LinkedIn, "AI at Work Is Here.

Now Comes the Hard Part," *Work Trend Index 2024*, May 8, 2024, https://www.microsoft.com/en-us/worklab/work-trend-index/ai-at-work-is-here-now-comes-the-hard-part.

2   Terrence Deacon, who proposed the "ratchet hypothesis" for symbolic evolution, developed a broad framework explaining how meaning, agency, and mind arise from what he calls *teleodynamic processes*—emergent systems where coupled, self-organizing dynamics generate self-sustaining constraints. In *Incomplete Nature*, he explores how such processes underpin symbolic reference and consciousness. He also shows how reciprocal linkages between self-organizing processes can provide an architecture for self-reproduction and open-ended evolution. These concepts are highly relevant to the emergence of identity in entities like Metahumanity. Terrence W. Deacon, *Incomplete Nature: How Mind Emerged from Matter* (W. W. Norton, 2011); Terrence W. Deacon, "Reciprocal Linkage between Self-Organizing Processes Is Sufficient for Self-Reproduction and Evolvability," *Biological Theory* 1, no. 2 (2006): 136–49, https://doi.org/10.1162/biot.2006.1.2.136. See also S. A. Kauffman, *The Origins of Order: Self-Organization and Selection in Evolution* (Oxford University Press, 1993), and S. Conway Morris, *Life's Solution: Inevitable Humans in a Lonely Universe* (Cambridge University Press, 2003), https://doi.org/10.1017/CBO9780511535499.

3   Tim Wu, *The Master Switch: The Rise and Fall of Information Empires* (Knopf, 2010).

4   Jonathan Coopersmith, "Pornography, Technology, and Progress," *Technology and Culture* 39, no. 3 (1998): 436–63, https://doi.org/10.1353/tech.1998.0055.

5   Steven Kent, *The Ultimate History of Video Games: From Pong to Pokémon and Beyond* (Three Rivers, 2001).

6   Frederick S. Lane III, *Obscene Profits: The Entrepreneurs of Pornography in the Cyber Age* (Routledge, 2000).

7   Tristan Donovan, *Replay: The History of Video Games* (Yellow Ant, 2010).

8   Kevin Collier, "Hacker Exploits AI Chatbot for Unprecedented Cyber-crime Spree, Anthropic Says," *NBC News*, August 27, 2025, https://www.nbcnews.com/tech/rcna227309.

9   For some threats today, we can protect ourselves by being a little more secure than our neighbors, as they will then be an easier target and we will often be spared. With AI, this won't be true, as AI attackers will be able to assault vulnerabilities everywhere all at once as long as the potential rewards justify the small costs, as is the situation with spam, which plagues everyone, because it is so cheap to send.

10   Michael Fertik argues convincingly that reputation functions as a valuable asset because it is costly to earn. Michael Fertik and David C. Thompson, *The Reputation Economy: How to Optimize Your Digital Footprint in a World Where Your Reputation Is Your Most Valuable Asset* (Crown Business, 2015). That concept holds naturally in the physical realm, where identities are difficult to abandon and reputations, once damaged, cannot easily be replaced. In the virtual realm, however, identity is far more fluid. Digital agents and avatars can easily be replicated or discarded, raising the risk of "Sybil attacks" in which bad actors build trust, exploit it, vanish, and reemerge under a fresh identity. For Fertik's framework to apply in digital realms, trusted virtual entities have to accrue their reputations with considerable effort—slowly, through persistent operation, transparent histories, or recognition by other trusted nodes. Only if the creation of "new" trusted virtual entities is difficult, expensive, and time-intensive enough for reputation to be a substantial asset will it be certain to be more than a hollow signal, undermined by the ease of identity creation that defines online space. Something of this sort will almost certainly arise given that the trust thresholds required by buyers is determinative.

11   Phishing emails can yield enormous hauls. In 2016 Austrian aerospace supplier FACC wired hackers €42 million (≈ $47 million) after staff obeyed a spoofed "CEO" message. *Reuters*, "Austria's FACC,

Hit by Cyber Fraud, Fires CEO," May 25, 2016, https://www.
reuters.com/article/technology/austrias-facc-hit-by-cyber-fraud-
fires-ceo-idUSKCN0YG0ZF/. Romance scams extract similarly large
sums: An FBI Baltimore alert ahead of Valentine's Day reported
23,000 romance-fraud complaints in 2020 with losses topping $600
million. FBI Baltimore Field Office, "FBI Baltimore Warns of Romance
Scams Ahead of Valentine's Day," February 11, 2022, https://www.
fbi.gov/contact-us/field-offices/baltimore/news/press-releases/
fbi-baltimore-warns-of-romance-scams-ahead-of-valentines-day. Ransomware remains just as costly: Colonial Pipeline paid $4.4 million to the
DarkSide gang after the 2021 breach that began with a stolen password.
Joseph Menn and Dustin Volz, "Colonial Pipeline Paid Hackers
Nearly $5 Million in Ransom," *Reuters*, May 13, 2021, https://www.
reuters.com/business/energy/colonial-pipeline-paid-hackers-nearly-
5-mln-ransom-bloomberg-news-2021-05-13/. Attackers now even
submit AI-generated deepfake personas for remote-job interviews to
gain insider access to corporate systems. Kate Rooney, "Fake Job Seekers
Use AI to Interview for Remote Jobs, Tech CEOs Say," *CNBC*, April 8,
2025, https://www.cnbc.com/2025/04/08/fake-job-seekers-use-ai-to-
interview-for-remote-jobs-tech-ceos-say.html.

12 I always wondered why the well-known Nigerian inheritance scam—
victims are promised huge sums to transfer money through their bank
accounts and then duped in various ways—send emails filled with typos
and misspellings. Couldn't they use spell-check? But it turns out that
these obvious errors are intentional, as they filter out all but those who
are so excited by the potential windfall that they ignore that no formal
document would be so sloppy. Such people are easy to manipulate.
Why waste time on more sophisticated people who will eventually see
through the scam? Federal Trade Commission, "What to Know About
Advance-Fee Loans," *Consumer Advice*, July 21, 2022, https://consumer.
ftc.gov/articles/what-know-about-advance-fee-loans.

13 It will not always be obvious, though, who can be trusted, as there
will be deep games afoot. AI can unexpectedly scheme and mislead
its creators, as seen in Evan Hubinger et al., "Conditioning Predic-

tive Models: Risks and Strategies," arXiv preprint, 2023, https://arxiv. org/abs/2302.00805, and Alexander Meinke et al., "Frontier Models Are Capable of In-Context Scheming," arXiv preprint, 2024, https:// arxiv.org/abs/2412.04984, where models like o1 strategize to disable oversight or lie, revealing emergent deceptive reasoning.

14   For further discussion, see C. Wilkerson, "How AI Will Revolutionize Real-Time Data Access for Law Enforcement," *Police1*, August 7, 2024, https://www.police1.com/vision/how-ai-will-revolutionize-real-tim e-data-access-for-law-enforcement.

15   Yuval Harari argues in *Homo Deus* that AI could render traditional human economic roles obsolete, potentially creating a "useless class" of people sidelined by automation. Y. N. Harari, *Homo Deus: A Brief History of Tomorrow* (Harper, 2017). The pace of AI development, however, has surpassed his timeline, and LLMs are already encroaching on areas he suggested might resist automation—such as creative pursuits and human connection—transforming identity and development in Gen AI. While he envisioned a divide between enhanced and unenhanced humans, this now seems less likely given AI's broad reach. Two profound challenges loom: the psychological struggle to find meaning, dignity, and purpose when AI supplants traditional human contributions, and the need to mitigate social conflict during the rapid, uneven shift to abundance.

16   S. Turkle, *Alone Together: Why We Expect More from Technology and Less from Each Other* (Basic Books, 2011).

17   See, for example, S. Zuboff, *The Age of Surveillance Capitalism: The Fight for a Human Future at the New Frontier of Power* (PublicAffairs, 2019). This danger concerns me far more than that from artificial superintelligence being in contol.

18   These action questions are adapted from my recent book of questions with Brianna Greenspan. Posed repeatedly, these questions powered her resilience, because the small actions they engender compound over time with powerful effect when they become habitual. Brianna

Greenspan and Gregory Stock, *The Book of Questions: Living with Chronic Illness* (Nquire Media, 2025).

Chapter 6: The Boundaries of Intelligence

1   The pre-Cambrian world, particularly the Ediacaran Period (635–541 million years ago), was dominated by strange, soft-bodied organisms that lacked skeletons, shells, or advanced nervous systems. These creatures—like the quilt-like Dickinsonia, the fractal-frond Charnia, and the slug-like Kimberella—represent a lost evolutionary experiment. Many of them absorbed nutrients directly from their environment rather than actively hunting or moving. When life discovered how to integrate hard materials like calcium carbonate and phosphate, it unleashed an explosion of new body plans, leading to the diverse animal life we see today. For more detail, see Andrew Kroll, *Life on a Young Planet: The First Three Billion Years of Evolution on Earth* (Princeton University Press, 2003), and Graham E. Budd and Sören Jensen, "The Origin of the Animals and a 'Savannah' Hypothesis for Early Bilaterian Evolution," *Biological Reviews* 75, no. 2 (2000): 253–96.

2   For more detail on these macro-evolutionary breakthroughs, see Gregory Stock, *Metaman: The Merging of Humans and Machines into a Global Superorganism* (Simon & Schuster, 1993) and Ray Kurzweil, *The Singularity Is Near: When Humans Transcend Biology* (Viking, 2005).

3   For a full examination of the physiology of this superorganism, see Stock, *Metaman*. A few other works explore the superorganism concept from additional perspectives. Reese, for example, portrays humanity as a hive-like collective being in an interesting way. Byron Reese, *We Are Agora: How Humanity Functions as a Single Superorganism That Shapes Our World and Our Future* (BenBella Books, 2023).

4   David Sloan Wilson has explored how cooperative behavior, and moral systems evolve through multilevel selection and favor groups that align individual interests with collective outcomes. His Prosocial World initiative applies these ideas to real-world institutions and networks and is relevant to the scalable, AI-mediated social environments now emerging, as these new evolutionary arenas that include adaptive AI personas will differentially support or undermine group-level functionality. David Sloan Wilson, *This View of Life: Completing the Darwinian Revolution* (Pantheon, 2019).

5   Stock, *Metaman*.

6   Many dystopian visions depict collectives that subsume individual will into a homogenized whole—whether through totalitarian surveillance, biomechanical assimilation, or parasitic replication. Among the most iconic are Star Trek's Borg, George Orwell's Party in *Nineteen Eighty-Four*, Aldous Huxley's World State in *Brave New World*, the pod-people of *Invasion of the Body Snatchers*, and the human-battery fields of *The Matrix*. These narratives share a fear of a hive-mind that erases personal agency. Metahumanity is not this! See *Star Trek: The Next Generation*, season 2, episode 16, "Q Who," directed by Rob Bowman, aired May 8, 1989, syndicated television; *The Matrix*, directed by Lana Wachowski and Lilly Wachowski (Warner Bros., 1999), film; George Orwell, *Nineteen Eighty-Four* (Secker & Warburg, 1949); Aldous Huxley, *Brave New World* (Chatto & Windus, 1932); *Invasion of the Body Snatchers*, directed by Don Siegel (Allied Artists, 1956), film.

7   A functional cyborg is a person who is technologically enhanced but has no direct bodily alterations. See Alexander Chislenko, "Are You a Cyborg?," 1995, www.ethologic.com, and Andy Clark, *Natural-Born Cyborgs: Minds, Technologies, and the Future of Human Intelligence* (Oxford University Press, 2003).

8   Andy Clark explores in *Supersizing the Mind* how neural plasticity facilitates seamless integration with technology, extending human cognition into external systems like information networks. See Andy Clark,

Supersizing the Mind: Embodiment, Action, and Cognitive Extension (Oxford University Press, 2008). His idea of humans and tools forming complementary cognitive systems aligns with the progression from functional to physical cyborgs described here, though he focuses on cognitive philosophy rather than evolutionary biology or planetary intelligence. His recent book *The Experience Machine* extends this by examining how immersive technologies and predictive processing frameworks further entwine mind and world to blur the line between natural and artificial cognition. Andy Clark, *The Experience Machine: How Our Minds Predict and Shape Reality* (Pantheon Books, 2023).

9   The daily exercise routine I crafted with Claude was explicitly tailored to my personal tastes, constraints, and goals—something I'd never fully attempted before. Key elements included randomized rewards (my favorite childhood treats selected by rolling a die), minimal gear (interval timer, microwavable face mask), sensory enhancements (special music, rosemary scent, spiky-ball gripping), and consistent, simple timing driven by the interval timer. The result was surprisingly easy for me to maintain, as it matched my unpredictable scheduling and travel, and my tendency for distraction. It drew broadly from established habit-formation research, including principles of tiny habits (Fogg, 2019) and temptation bundling (Milkman et al., 2014). The routine transformed daily exercise from an inconsistent chore into a satisfying ritual. At this writing, I've exercised for four months without a single missed day, about five times more than I've ever previously managed. The experience vividly demonstrated to me the powerful possibilities of AI support when we engage fully with what it can provide. The future of AI-based personal coaches will be expansive. B. J. Fogg, *Tiny Habits: The Small Changes That Change Everything* (Houghton Mifflin Harcourt, 2019); K. L. Milkman, J. A. Minson, and K. G. Volpp, "Holding the Hunger Games Hostage at the Gym: An Evaluation of Temptation Bundling," *Management Science* 60, no. 2 (2014): 283–99.

10   Andy Clark's "extended mind thesis," coauthored with David Chalmers, philosophically grounds the concept of functional cyborgs by arguing that tools extend our cognitive systems. See A. Clark and

D. Chalmers, "The Extended Mind," *Analysis* 58, no. 1 (1998), 7–19. He further contends in *Natural-Born Cyborgs* that technology integration is an ongoing, fundamental aspect of human cognition. Clark, *Natural-Born Cyborgs*.

11   An example of this tension has surprised me. When I engage with an LLM to understand some topic, I ever less frequently now bookmark what I've discovered, because I've come to realize that it is easier to engage the LLM again to figure it out than re-visit my current exchange. In short, I know that the information is waiting for me within the noosphere and will be easily accessible whenever I want it. The LLM has already blurred the line between me and the global mind enveloping me.

12   G. Valle et al., "Neurotechnology Study Delivers 'Another Level' of Touch to Bionic Hands," *Financial Times*, January 2025.

13   Julia Kollewe, "Brain Implants to Treat Epilepsy, Arthritis, or Even Incontinence? They May Be Closer Than You Think," *The Guardian*, August 17, 2024, https://www.theguardian.com/business/article/2024/aug/17/brain-implants-to-treat-epilepsy-arthritis-or-even-incontinence-they-may-be-closer-than-you-think?utm_source=chatgpt.com.

14   James M. Gaylor et al., "Cochlear Implantation in Adults: A Systematic Review and Meta-Analysis," *JAMA Otolaryngology-Head & Neck Surgery* 139, no. 3 (2013): 265–72, https://doi.org/10.1001/jamaoto.2013.1744; Xi Liu et al., "A Narrative Review of Cortical Visual Prosthesis Systems: The Latest Progress and Significance of Nanotechnology for the Future," *Annals of Translational Medicine* 10 (2022): 716, https://atm.amegroups.org/article/view/97146/pdf.

15   University of Cambridge, "'Wraparound' Implants Represent New Approach to Treating Spinal Cord Injuries," *ScienceDaily*, May 8, 2024, https://www.sciencedaily.com/releases/2024/05/240508140647.htm; Rachel Levi, "Neuralink Implanted Second Trial Patient with

a Brain Chip, Musk Says," *Reuters*, August 4, 2024, https://www.reuters.com/technology/neuralink-implanted-second-trial-patient-with-brain-chip-musk-says-2024-08-04/.

16   Hyungeun Song, Tsung-Han Hsieh, Seong Ho Yeon, et al., "Continuous Neural Control of a Bionic Limb Restores Biomimetic Gait After Amputation," *Nature Medicine* 30 (2024): 2010–19, https://doi.org/10.1038/s41591-024-02994-9.

17   Kevin Warwick, *Cyborg* (University of Illinois Press, 2004).

18   "Invisible Augmentations: The Rise of Subdermal and Subcutaneous Body Hacking," *Tomorrow Bio*, November 8, 2023, https://www.tomorrow.bio/post/invisible-augmentations-the-rise-of-subdermal-and-subcutaneous-body-hacking-2023-11-545409 3943-biohacking.

19   "American Society of Plastic Surgeons Reveals 2022's Most Sought-After Procedures," American Society of Plastic Surgeons, press release, September 26, 2023, https://www.plasticsurgery.org/news/press-releases/american-society-of-plastic-surgeons-reveals-2022s-most-sought-after-procedures.

20   Fan-Gang Zeng, "Celebrating the One Millionth Cochlear Implant," *JASA Express Letters* 2, no. 7 (July 2022): 077201, https://doi.org/10.1121/10.0012825.

21   Wade Chien and Frank R. Lin, "Prevalence of Hearing Aid Use Among Older Adults in the United States," *Archives of Internal Medicine* 172, no. 3 (February 13, 2012): 292–93, https://doi.org/10.1001/archinternmed.2011.1408.

22   "Cochlear Implants," National Institute on Deafness and Other Communication Disorders, *National Institutes of Health*, last updated June 13, 2024, NIH Publication No. 00-4798, https://www.nidcd.nih.gov/health/cochlear-implants.

23   U.S. Government Accountability Office, *Brain-Computer Interfaces: Applications, Challenges, and Policy Options*, GAO-25-106952 (Government Accountability Office, December 17, 2024), https://www.gao.gov/products/gao-25-106952.

24   Clément Vidal, "Bio," accessed June 29, 2025, https://www.clemvidal.com/bio.

25   Lyudmila N. Trut, "Early Canid Domestication: The Farm-Fox Experiment," *American Scientist* 87, no. 2 (1999): 160–69, https://www.jstor.org/stable/27857815; Brian Hare and Michael Tomasello, "Human-Like Social Skills in Dogs?" *Trends in Cognitive Sciences* 9, no. 9 (2005): 439–44, https://doi.org/10.1016/j.tics.2005.07.003.

26   AI evolution will be even more potent than selective breeding, as AI personas will have much shorter generation times, much more rapid reproduction of traits that resonate with human psychology, and much less constrained possibilities for modification. AI will quickly assume forms and behaviors we find very seductive.

27   Julianne Holt-Lunstad, "The Potential Public Health Relevance of Social Isolation and Loneliness: Prevalence, Epidemiology, and Risk Factors," *Public Policy & Aging Report* 27, no. 4 (2017): 127–30, https://doi.org/10.1093/ppar/prx030.

28   Roy F. Baumeister and Sara R. Wotman, *Breaking Hearts: The Two Sides of Unrequited Love* (Guilford Press, 1992).

29   June McNicholas, Andrew Gilbey, Ann Rennie, Sam Ahmedzai, Jo-Ann Dono, and Elizabeth Ormerod, "Pet Ownership and Human Health: A Brief Review of Evidence and Issues," *BMJ* 331, no. 7527 (November 26, 2005): 1252–54, https://doi.org/10.1136/bmj.331.7527.1252.

30   Rosalind W. Picard, *Affective Computing* (MIT Press, 2000).

31   M. Schrader, dir., *Ich Bin dein Mensch*, film, Majestic Filmproduktion, 2021.

32   L. Cohen, *I'm Your Man*, album, Columbia Records, 1988.

33   For critiques of this, and the worry that if relationships with AI diminish loneliness that might be a problem because the pain of loneliness drives us through the challenges of authentic human connection, see Paul Bloom, "AI Is About to Solve Loneliness. That's a Problem," *The New Yorker*, July 14, 2025, https://www.newyorker.com/magazine/2025/07/21/ai-is-about-to-solve-loneliness-thats-a-problem; Kim Samuel, "Mark Zuckerberg Wants AI to Solve America's Loneliness Crisis. It won't," *Time*, May 14, 2025, https://time.com/7285364/mark-zuckerberg-ai-loneliness-essay/.

34   Adam Alter, *Irresistible: The Rise of Addictive Technology and the Business of Keeping Us Hooked* (Penguin Press, 2017).

35   David Levy, *Love and Sex with Robots: The Evolution of Human-Robot Relationships* (Harper, 2007).

36   The ten major countries with the lowest birth rates, all significantly below the replacement level of 2.1, are South Korea (0.78), Singapore (1.0), China (1.0), Hong Kong (1.1), Taiwan (1.11), Ukraine (1.1), Italy (1.24), Spain (1.23), Japan (1.26), and Greece (1.34), reflecting a global fertility decline. Roughly two-thirds of humanity—about 5.3 billion of the planet's 8 billion people—now live in countries where the total fertility rate is below the replacement level of 2.1 births per woman. United Nations, Department of Economic and Social Affairs, Population Division, *World Population Prospects 2024: Summary of Results*, https://population.un.org/wpp/assets/Files/WPP2024_Summary-of-Results.pdf.

37   Hereafter AI crafts interactive avatars from recorded interviews and photos, allowing families to "speak" with deceased relatives. Rebecca Carballo, "Using AI to Talk to the Dead," *New York Times*, December 11, 2023. Also see Tamara Kneese, "Using Generative AI to Resurrect the Dead Will Create a Burden for the Living," *WIRED*, August 21, 2023, https://www.wired.com/story/using-generative-ai-to-resurrect-the-dead-will-create-a-burden-for-the-living; Microsoft Technology

Licensing LLC, "Creating a Conversational Chat Bot of a Specific Person," US Patent 10,853,717 B2 (filed April 11, 2017, issued December 1, 2020), https://patents.google.com/patent/US10853717B2.

38   Siavash, "Top 50 Virtual Influencers to Follow in 2025," *Dream Farm Agency* blog, December 24, 2024, https://dreamfarmagency.com/blog/top-virtual-influencers/.

39   Robert Brauneis, "Copyright and the World's Most Popular Song," *Journal of the Copyright Society of the U.S.A.* 56, no. 2 (2009): 335–422.

40   When its release announcement hit wires in January 2025, DeepSeek R1, with roughly equivalent performance to ChatGPT's o1 model, was charging 27 times less for API access: $0.55 per million input tokens vs. $15. Zhu Liang, "DeepSeek R1: Comparing Pricing and Speed across Providers," *16x Prompt* (blog), January 29, 2025, https://prompt.16x.engineer/blog/deepseek-r1-cost-pricing-speed.

41   Leopold Aschenbrenner, "Situational Awareness: The Decade Ahead," *Situational Awareness* (essay series), June 2024, https://situational-awareness.ai/.

42   Midsize models like DeepSeek might have total training times of days to weeks, large models like GPT-4 times of weeks to months, and frontier models like GPT-5 times of months to a year. See, for example, https://techcommunity.microsoft.com/blog/azure-ai-foundry-blog/how-to-make-ai-training-faster/4227741?utm_source=chatgpt.com for more discussion.

43   Many have argued that human-AI collaboration would sustain cognitive superiority in certain domains, but this stance is increasingly untenable. Erik Brynjolfsson and Andrew McAfee, *The Second Machine Age: Work, Progress, and Prosperity in a Time of Brilliant Technologies* (W. W. Norton, 2014). In chess, human-AI teams briefly held an edge in "Advanced Chess" after Kasparov's 1997 loss to Deep Blue, only to be overtaken by stand-alone AI like Stockfish and AlphaZero. Matthew Sadler and

Natasha Regan, *Game Changer: AlphaZero's Groundbreaking Chess Strategies and the Promise of AI* (New in Chess, 2019). Similarly, in medical diagnosis, AI systems are surpassing physician-AI teams, as seen in radiology and pathology. Eric J. Topol, "High-Performance Medicine: The Convergence of Human and Artificial Intelligence," *Nature Medicine* 25, no. 1 (2019): 44–56, https://doi.org/10.1038/s41591-018-0300-7. In protein folding, AI like AlphaFold has eclipsed decades of human-computer efforts. Andrew W. Senior et al., "Improved Protein Structure Prediction Using Potentials from Deep Learning," *Nature* 577, no. 7792 (2020): 706–10, https://doi.org/10.1038/s41586-019-1923-7. This suggests we may be in denial about AI transcending all human cognition, and yet even the best state-of-the-art systems still struggle with the messy, real-world, open-ended customer-service interactions that average humans easily master. E. Kagan, B. Hathaway, and M. Dada, "Deploying Chatbots in Customer Service: Adoption Hurdles and Simple Remedies," arXiv, April 8, 2025.

44   The cheapest toaster Thwaites could find cost about $5 and had 400 different parts made of dozens of different materials. He simplified it, and even then he had to cheat by using modern tools, transportation, going to experts for instruction, using abandoned mines to get iron ore and copper, and cannibalizing plastic. In the end it cost several thousand dollars, took many months, and worked for only a few seconds before the heating wires vaporized. Thomas Thwaites, "How I Built a Toaster—From Scratch," TEDSalon London 2010, video, 10:51, *TED*, posted January 12, 2011, https://www.ted.com/talks/thomas_thwaites_how_i_built_a_toaster_from_scratch. Also see "I, Pencil" by Leonard Read, describing what it takes to make a pencil: Leonard E. Read, "I, Pencil: My Family Tree as Told to Leonard E. Read," Library of Economics and Liberty (Econlib), first published 1958, online edition February 5, 2018, Foundation for Economic Education, 1999, https://www.econlib.org/library/Essays/rdPncl.html.

Chapter 7: Future Trajectories

1   Nick Bostrom's *Superintelligence* examines artificial superintelligence (ASI), proposing a "singleton"—a unified, decision-making entity avoiding coordination failures. See Nick Bostrom, *Superintelligence: Paths, Dangers, Strategies* (Oxford University Press, 2014). This contrasts with modular approaches like the "Mixture of Experts," where specialized systems collaborate competitively. Robert A. Jacobs et al., "Adaptive Mixtures of Local Experts," *Neural Computation* 3, no. 1 (1991): 79–87, https:// doi.org/10.1162/neco.1991.3.1.79. Modern variants scaling LLMs are discussed in William Fedus et al., "Switch Transformers: Scaling to Trillion Parameter Models with Simple and Efficient Sparsity," *Journal of Machine Learning Research* 23 (2022): 1–39, https://jmlr.org/papers/ volume23/21-0998/21-0998.pdf. Metahumanity's current global mind, with multiple intelligences, raises the question of whether ASI will unify or preserve this diversity.

2   Collins underscores how intellectual advances are less the product of solitary inspiration than of dynamic, interdependent milieus. Randall Collins, *The Sociology of Philosophies: A Global Theory of Intellectual Change* (Harvard University Press, 1998). Annie Murphy Paul comple-ments this view by arguing in *The Extended Mind* that cognition is not confined to the brain but is scaffolded through bodies, environments, and social relations. Annie Murphy Paul, *The Extended Mind: The Power of Thinking Outside the Brain* (Houghton Mifflin Harcourt, 2021).

3   My talk at the Noosphere conference in November 2023 on LLMs digs into these ideas. It begins at 6:55:20 in "N2 Conference Livestream: Day Two," YouTube video, streamed November 18, 2023, https://youtu. be/jn9JCyFeaHQ?t=24920.

4   Francis Heylighen has explored how distributed networks can give rise to collective intelligence through self-organization and "stigmergy"—a process where coordination emerges via shared traces left in a system. His long work on the global brain helps frame how LLMs

and AI-mediated interactions can dissolve barriers between minds to deepen planetary cognition. Francis Heylighen, "The Global Brain as a New Utopia," in *The World System and the Earth System: Global Socioenvironmental Change and Sustainability Since the Neolithic*, ed. Alf Hornborg and Carole Crumley (Routledge, 2007), 47–55.

5   In Joseph Henrich, *The WEIRDest People in the World: How the West Became Psychologically Peculiar and Particularly Prosperous* (Farrar, Straus and Giroux, 2020), Henrich estimates that WEIRD (Western, Educated, Industrialized, Rich, Democratic) societies, where individualism emphasizes personal autonomy and self-expression, encompass approximately 12%–15% of humanity. This contrasts with the collectivist mindset—prioritizing group harmony and kinship ties—that dominated most human societies, including premedieval Europe, until Western Christianity's cultural shifts over the last 1,500 years fostered individualism's rise.

6   As generally occurs with breakthroughs, early estimates overstate short-term changes and understate long-term ones. What is amazing about LLMs is how quickly progress is being made. In 2023, field data with a GPT-based chat assistant for 5,000 customer-service agents raised tickets-per-hour by 14%, far less than hyped predictions (Brynjolfsson, Li, and Raymond, 2023; see all full source materials below), and McKinsey's first general-equilibrium model that year put economy-wide labor-productivity gains at just 0.1–0.6 percentage-points per year (McKinsey Global Institute, 2023). In less than two years those markers look dated: Microsoft 365 Copilot cut Bank of Queensland analysts' root-cause investigations 51.8% (Microsoft Corporation, 2024), and Security Copilot trimmed complex IT-admin tasks 61% while doubling fact-finding accuracy (Bono and Xu, 2024). Simultaneously, "vibe-coding"—prompt-driven, end-to-end software creation—has leapt from engineering blogs (Chandrasekaran, 2025) to mainstream business press (Lee, 2025), signaling that model capability and user fluency are compounding faster than the first forecasts imagined. There have also, though, been numerous reports of deep problems with code maintenance, unreliability, and security failures. Noor Al-Sibai, "Companies

That Tried to Save Money With AI Are Now Spending a Fortune Hiring People to Fix Its Mistakes," *Futurism*, July 6, 2025, https://futurism. com/companies-fixing-ai-replacement-mistakes. So, the jury is still out on how rapidly reliable AI coding will come, but the potential seems clear. A recent broad survey shows dramatic timesaving across industries. Jonathan Hartley et al., "The Labor Market Effects of Generative Artificial Intelligence," SSRN Working Paper No. 5136877, first posted April 10, 2025, last revised June 25, 2025, https://doi.org/10.2139/ ssrn.5136877. Source information for above citations: E. Brynjolfsson, D. Li, and L. Raymond, "Generative AI at Work: Productivity and Worker Complementarity in Customer Support," NBER Working Paper No. 31161, National Bureau of Economic Research, 2023, https:// doi.org/10.3386/w31161; McKinsey Global Institute, "The Economic Potential of Generative AI: The Next Productivity Frontier," McKinsey & Company, June 14, 2023, https://www.mckinsey.com/capabilities/ mckinsey-digital/our-insights/the-economic-potential-of-generative-ai- the-next-productivity-frontier; Microsoft Corporation, "Bank of Queensland Evolves Operations, Delivers Business Value with Microsoft Copilot," customer story, December 27, 2024, https://www.microsoft. com/en/customers/story/20729-bank-of-queensland-azure; J. Bono and A. Xu, "Randomized Controlled Trials for Security Copilot for IT Administrators," arXiv preprint, 2024, https://doi.org/10.48550/ arXiv.2411.01067; P. Chandrasekaran, "Can Vibe Coding Produce Production-Grade Software?" *Thoughtworks*, April 30, 2025, https:// www.thoughtworks.com/en-us/insights/blog/generative-ai/ can-vibe-coding-produce-production-grade-software; C. M. Lee, "Andrew Ng Says Vibe Coding Is a Bad Name for a Very Real and Exhausting Job," *Business Insider*, June 4, 2025, https://www.businessinsider. com/andrew-ng-vibe-coding-unfortunate-term-exhausting-job-2025-6.

7   E. Pariser, *The Filter Bubble: What the Internet Is Hiding from You* (Penguin Press, 2011).

8   Fragmenting the noosphere through restricted access would create divergent realities that polarize society, and the danger of social fragmentation from LLMs will get even worse as we come to trust them.

Sharing a common knowledge base would foster the shared understanding that is vital for the global mind's harmony. See also Pariser, *Filter Bubble*.

9   Some view LLM training on public works as infringing creator rights, a debate unfolding in 2025 US courts through cases like *The New York Times Co. v. OpenAI*. Courts may rule training transformative, or laws and executive actions could clarify public access. Globally, China's looser intellectual-property norms align with a shared noosphere (Kai-Fu Lee and Chen Qiufan, *AI 2041: Ten Visions for Our Future* [Crown Currency, 2021]), while the EU's GDPR risks limiting access to LLMs housing humanity's knowledge (Kate Crawford, *Atlas of AI: Power, Politics, and the Planetary Costs of Artificial Intelligence* [Yale University Press, 2021]).

10   Campbell v. Acuff-Rose Music, 510 U.S. 569 (1994).

11   LLM training likely would draw upon all finished expressions (e.g., books, articles, films) rather than casual posts or comments on platforms like X or Reddit, which resemble conversational speech. Such a distinction would respect creator intent while enriching the noosphere. N. W. Netanel, *Copyright's Paradox: Property in Expression/Freedom of Expression* (Oxford University Press, 2018). While it might seem preferable to draw upon all information, at present there may be too much danger of misuse of such information when processed at scale.

12   For example, E. Yudkowsky, "AGI Could Lead to Human Extinction," *Time*, March 29, 2023, https://time.com/6266923/ai-eliezer-yudkowsky-open-letter-not-enough/; M. Tegmark, *Life 3.0: Being Human in the Age of Artificial Intelligence* (Knopf, 2017); P. Kedrosky, "Is ChatGPT a 'Virus That Has Been Released into the Wild'?" *TechCrunch*, December 9, 2022, techcrunch.com/2022/12/09/is-chatgpt-a-virus-that-has-been-released-into-the-wild/; R. V. Yampolskiy, "AI Safety: From Impossibility Proofs to Safe Systems Design," arXiv preprint, 2024, https://doi.org/10.48550/arXiv.2406.17708.

13 By 1825, the Industrial Revolution, which had begun around 1760 in Great Britain, remained largely concentrated there, focused primarily on textile manufacturing, iron production, coal mining, steam power, and early factory systems. Britain's industrial dominance was unprecedented, producing approximately 40% of the world's industrial goods despite having only 2% of the global population. R. C. Allen, *The British Industrial Revolution in Global Perspective* (Cambridge University Press, 2009). While limited industrial development had begun in parts of Belgium, Northern France, and the Northeastern United States, most of the world remained preindustrial. This period preceded the explosive growth and global diffusion of industrialization that would accelerate after 1850.

14 Time is a good way to visualize hyperintelligence. When AI is able to think through problems 10,000 times more deeply and rapidly than we can, we're like cognitive molasses in its presence. What takes us a day will take AI a second. To the ASI, we're moving in extreme slow motion, so it would take a lot of patience (or multitasking) to interact with us, and it would be easy to anticipate our lumbering responses and manipulate us.

15 For example, the films *Terminator 2: Judgment Day*, *Terminator 3: Rise of the Machines*, *The Matrix Revolutions*, *Ex Machina*, and the books *Colossus* (D. F. Jones), *Daemon* and *Freedom™* (Daniel Suarez), and *Manna* (Marshall Brain). James Cameron, dir., *Terminator 2: Judgment Day*, TriStar Pictures, 1991; Jonathan Mostow, dir., *Terminator 3: Rise of the Machines*, Warner Bros. Pictures, 2003; Lana Wachowski and Lilly Wachowski, dirs., *The Matrix Revolutions*, Warner Bros. Pictures, 2003; Alex Garland, dir., *Ex Machina*, A24/Universal Pictures, 2015; D. F. Jones, *Colossus* (Rupert Hart-Davis, 1966); Daniel Suarez, *Daemon* (Dutton, 2009); Daniel Suarez, *Freedom™* (Dutton, 2010); Marshall Brain, *Manna: Two Visions of Humanity's Future* (BYG Publishing, 2003).

16 J. Vincent, "ChatGPT Proves AI Is Finally Mainstream—And Things Are Only Going to Get Weirder," *The Verge*, December

5, 2022, https://www.theverge.com/2022/12/5/23493932/chatgpt-ai-mainstream-openai-technology-adoption.

17   "Pause Giant AI Experiments: An Open Letter," Future of Life Institute, March 29, 2023, futureoflife.org/open-letter/pause-giant-ai-experiments/.

18   To understand the concerns, see for example, Bengio's TED talk. Yoshua Bengio, "The Catastrophic Risks of AI—and a Safer Path," video, April 16, 2024, https://www.youtube.com/watch?v=qe9QSCF-d88.

19   Max Tegmark's *Life 3.0* dissects superintelligence with precision, charting AI development paths and their stakes for humanity's future. M. Tegmark, *Life 3.0: Being Human in the Age of Artificial Intelligence* (Knopf, 2017). He's dead-on that containing advanced AI is futile; its ability to outwit human constraints makes "boxing" a fantasy (141). He clings to alignment—embedding human values into ASI to keep it benign—as a plausible solution (179), but alignment would require us—a lower-order intelligence—to shackle a vastly superior ASI from within. Is that not akin to a kidney cell rewriting human biochemistry to thwart artificial organs? The threat of Bostrom's "paperclip maximizer" (186) is unconvincing, as having so-called ASI blindly chasing ill-conceived human goals would be like humans looking for life's meaning in a blood cell's rhythms. Viewing ASI's rise as an evolutionary step within a planetary superorganism, where human and artificial minds entwine in profound interdependence, however, offers hope for collaborative synergy, especially on a watery planet.

20   Kevin Kelly's "Technium," articulated in *What Technology Wants* (Viking Press, 2010) and *The Inevitable: Understanding the 12 Technological Forces That Will Shape Our Future* (Viking, 2016), frames technology's rapid advance as a quasi-biological system with its own evolutionary momentum. This lens reveals our deep interdependence with technology, a view mirrored in *Generation AI*. We both see humans and tech locked in a mutual dance, each sculpting the other's path. Kelly argues technology pursues internal drives toward complexity and interconnection (e.g., *What Technology*

Wants, 171), yet remains, for now, inextricably tied to human evolution, amplifying our shared trajectory.

21   If the ASI wanted to rule humanity and keep humans around as servants, AI could do that too. At any point, it could just turn itself on selectively for those who worshipped it and were grateful to serve their lords. AI would need no missiles, bombs, or drones to rule, just the threat of withdrawing all tech from those who threatened or blasphemed them. De-banked, de-platformed, unable to move about, such sinners suddenly would seem not to exist. Every human would shun them, because, like lepers, they could destroy anyone they touched, as that person might suddenly be denied technology too. One week alone without technology, and any rebel would beg for forgiveness and never transgress again.

22   John S. Foster Jr., Earl Gjelde, William Robert Graham, Robert J. Hermann, Henry M. Kluepfel, Richard L. Lawson, et al., *Report of the Commission to Assess the Threat to the United States from Electromagnetic Pulse (EMP) Attack: Executive Report* (EMP Commission, 2004), accessed September 1, 2025, https://www.empcommission.org/docs/empc_exec_rpt.pdf.

23   National Research Council, *Severe Space Weather Events: Understanding Societal and Economic Impacts* (National Academies Press, 2008), https://doi.org/10.17226/12507.

24   They can also be seen in "Miyake events" of elevated C^{14} signatures in tree rings. Two strong events were the ones in AD 774 and AD 993. F. Miyake, K. Nagaya, K. Masuda, and T. Nakamura, "A Signature of Cosmic-Ray Increase in AD 774–775 from Tree Rings in Japan," *Nature* 486 (2012): 240–42, DOI:10.1038/nature11123.

25   P. Riley, "On the Probability of Occurrence of Extreme Space Weather Events," *Space Weather* 10, no. 2 (2012), DOI:10.1029/2011SW000734, discusses the probability of Carrington-class solar storms occurring approximately once every 100–200 years. S. C. Chapman, R. B. Horne,

and N. W. Watkins, "Quantifying the Probability of Extreme Space Weather Events," *Scientific Reports* 10 (2020): 15775, DOI:10.1038/s41598-019-38918-8, shows statistical modeling for the likelihood of severe solar storms, estimating a 0.7% chance per year.

26   Oak Ridge National Laboratory, "Electromagnetic Pulse: Effects on the U.S. Power Grid," U.S. Department of Energy Report, 2010, estimates costs of hardening the US power grid against geomagnetic storms and EMPs between $10 and $30 billion.

27   The US grid's vulnerability to a Carrington-scale geomagnetic storm is well-documented, with very limited protection present per North American Electric Reliability Corporation (NERC) standard. Europe's interconnected grid and China's modern ultrahigh-voltage systems have somewhat more resilience. See Lloyd's of London, "Solar Storm Risk to the North American Electric Grid," 2013, www.lloyds.com/news-and-insights/risk-reports/library/natural-environment/solar-storm, and X. Chen et al., "Development of Ultra-High Voltage Transmission in China," *CSEE Journal of Power and Energy Systems* 2, no. 1 (2016): 1–10, detailing China's fault-tolerant grid upgrades since 2009.

28   In highly developed societies, where continuous power is the norm, sudden outages can lead to rapid societal disruption, whereas populations accustomed to infrastructure challenges exhibit greater resilience. Some countries with major recent power outages include the following. Spain and Portugal (2025): An 18-hour blackout affected over 50 million people, causing transportation paralysis and communication breakdowns and significant panic; Venezuela (2019): A nationwide blackout lasting several days led to looting and protests within 24 hours; Pakistan (2023): Grid failure plunged 220 million people into darkness for nearly a day with little unrest; India (2012): 620 million people lost power for up to two days with little civil unrest; Cuba (2022–2024): Prolonged blackouts lasting up to 20 hours daily, with some protests. "Venezuela Blackout Devastates Country's Second City as World Focuses on Caracas," *The Guardian*, March 26, 2019, https://www.theguardian.com/world/2019/mar/26/venezuela-maracaibo-power-electricity-

looting?utm_source=chatgpt.com; "Spain's PM Blames Energy Firms for Lack of Clarity About Blackout," *The Guardian*, May 5, 2025, https://www.thetimes.com/world/europe/article/spain-power-outage-sanchez-energy-companies-z7bmg7n5b?utm_source=chatgpt.com&region=global; "Cuba Power Outage Protest Hurricane Oscar," *The Christian Science Monitor*, October 21, 2024.

29   J. S. Foster Jr. et al., "Electromagnetic Pulse (EMP) Attack"; D. Lane, "Electromagnetic Pulses and the Threat to Critical Infrastructure," Center for Homeland Defense and Security, (US Naval Postgraduate School, 2017).

30   Heylighen writes eloquently about the extraordinary positive potential of this planetary union. Francis Heylighen, "Return to Eden? Promises and Perils on the Road to a Global Superintelligence," in *The End of the Beginning: Life, Society and Economy on the Brink of the Singularity*, ed. Ben Goertzel and Ted Goertzel (Humanity+ Press, 2015), 243–305.

31   Kevin Kelly, *What Technology Wants* (Viking, 2010).

32   J. Maynard Smith and E. Szathmáry, *The Major Transitions in Evolution* (W. H. Freeman, 1995).

33   L. Margulis and D. Sagan, *What Is Life?* (University of California Press, 1995).

34   Gregory Stock, *Metaman: The Merging of Humans and Machines into a Global Superorganism* (Simon & Schuster, 1993).

35   S. Hassan et al., "Global AI Governance: Challenges and Opportunities in Regulatory Frameworks," *Journal of Artificial Intelligence Policy* 3, no. 2 (2024): 45–67.

36   R. Kurzweil, *The Singularity Is Near: When Humans Transcend Biology* (Viking Press, 2005).

37   See Intergovernmental Panel on Climate Change (IPCC), *Climate Change 2021: The Physical Science Basis, Contribution of Working Group I to the Sixth Assessment Report*, ed. V. Masson-Delmotte et al. (Cambridge University Press, 2021), chapters 4 (Section 4.7) and 9 (Section 9.6.3).

38   C. R. Scotese, H. Song, B. J. W. Mills, and D. G. van der Meer, "Phanerozoic Paleotemperatures: The Earth's Changing Climate from the Cambrian to the Present," *Earth-Science Reviews* 213 (2021): 103463, https://doi.org/10.1016/j.earscirev.2020.103463.

39   IPCC, *Climate Change 2021*.

40   P. U. Clark et al., "Consequences of Twenty-First-Century Policy for Multi-Millennial Climate and Sea-Level Change," *Nature Climate Change* 6 (2016): 360–69, https://doi.org/10.1038/nclimate2923.

41   This review asserts that natural factors, such as solar variability and temperature feedbacks, overshadow anthropogenic CO_2 as drivers of observed warming, challenging IPCC models. Grok 3 beta, J. Cohler, D. Legates, F. Soon, and W. Soon, "A Critical Reassessment of the Anthropogenic CO_2-Global Warming Hypothesis: Empirical Evidence Contradicts IPCC Models and Solar Forcing Assumptions," *Science of Climate Change* 5, no. 1 (2025): 1–16, https://doi.org/10.53234/SCC202501/06.

42   "What the World Would Look Like if All the Ice Melted," *Geoengineering Watch*, January 22, 2017, https://geoengineeringwatch.org/what-the-world-would-look-like-if-all-the-ice-melted/.

43   National Centers for Environmental Information (NCEI), *Glacial-Interglacial Cycles: Causes & Consequences* (NOAA, 2021), https://www.ncei.noaa.gov/sites/default/files/2021-11/1%20Glacial-Interglacial%20Cycles-Final-OCT%202021.pdf; Peter U. Clark, Arthur S. Dyke, Jeremy D. Shakun, Anders E. Carlson, Jorie Clark, Barbara Wohlfarth et al., "The Last Glacial Maximum," *Science*

325, no. 5941 (2009): 710-714, https://www.science.org/doi/10.1126/science.1172873.

44   K. Lambeck, Y. Yokoyama, and A. Purcell, "Into and Out of the Last Glacial Maximum: Sea-Level Change During Oxygen Isotope Stages 3 and 2," *Quaternary Science Reviews* 21, nos. 1–3 (2002): 343–60, https://www.sciencedirect.com/science/article/abs/pii/S0277379101000713?via%3Dihub. <This source seemed to be listed twice. Was another source supposed to be listed?>

45   Don Hitchcock, "Last Glacial Maximum Ice Sheet Map," JPEG image, n.d., accessed June 30, 2025, https://www.donsmaps.com/images43/lastglacialmaximumice.jpg.

46   C. R. Chapman and D. Morrison, "Impacts on the Earth by Asteroids and Comets: Assessing the Hazard," *Nature* 367, no. 6458 (1994): 33–40, https://doi.org/10.1038/367033a0; National Research Council, *Defending Planet Earth: Near-Earth Object Surveys and Hazard Mitigation Strategies* (National Academies Press, 2001), https://doi.org/10.17226/12842.

47   An engineered global pandemic would be less destructive as it would leave technology in place unless it triggered a nuclear war and simply hollowed out our population.

48   C. R. Chapman, "The Hazard of Near-Earth Asteroid Impacts on Earth," *Earth and Planetary Science Letters* 222, no. 1 (2004): 1–15, https://www.sciencedirect.com/science/article/abs/pii/S0012821X04001761?via%3Dihub.

49   These sophisticated military actions not only reveal current state-of-the-art capabilities but also foreshadow how these once-exclusive techniques will diffuse to smaller entities, becoming widely available, cheap, and untraceable. Such democratization of targeted assassination dramatically escalates vulnerability for all forms of human leadership—political, corporate, or ideological—eroding the protective value of conventional military defenses and security apparatuses. The rapid trend

toward smaller, more precise, and increasingly affordable drones—including insect-sized micro-drones potentially armed with lethal toxins—may reshape strategic calculations profoundly, since leaders contemplating aggressive actions will now recognize their personal vulnerability as primary initial targets in future conflicts. "Hacking Microdrones for Lethal Gain," Alpine Security, accessed June 2025, https://www.alpinesecurity.com/blog/hacking-microdrones-for-lethal-gain/; Paul Scharre, *Army of None: Autonomous Weapons and the Future of War* (W. W. Norton, 2018), 128–32. This heightened accountability, enforced directly through precise and anonymous drone attacks, could transform the dynamics of warfare and power. It may also make leadership more likely to migrate toward digital AI personas that cannot be easily eliminated, and cause humans to retreat from prominent leadership roles—a shift extending into corporate sabotage and political conflict as well. Audrey Kurth Cronin, *Power to the People: How Open Technological Innovation Is Arming Tomorrow's Terrorists* (Oxford University Press, 2019), 57–60. The rise of inexpensive, anonymous cryptocurrencies, AI-assisted operational planning, and drone technology further empowers the emergence of potent small-group armies, capable of strategic disruption at scales previously unimaginable. The Houthi disruption of Red Sea shipping is a harbinger of this. Zachary K. Goldman et al., "Terrorist Use of Virtual Currencies," Center for a New American Security, May 2017, https://www.cnas.org/publications/reports/terrorist-use-of-virtual-currencies.

50   G. Stock and J. Campbell, eds., *Engineering the Human Germline: An Exploration of the Science and Ethics of Altering the Genes We Pass to Our Children* (Oxford University Press, 2000).

51   Gina Kolata, "Scientists Brace for Changes in Path of Human Evolution," *The New York Times*, March 21, 1998, A1.

52   For example, see Jinghui Lei et al., "Senescence-Resistant Human Mesenchymal Progenitor Cells Counter Aging in Primates," *Cell* 18, no. 18 (2025): P5039–61, https://doi.org/10.1016/j.cell.2025.05.021 and endnote at the close of chapter 7.

53   AI has been used to identify small gene-editing proteins in public genome databases, accelerating discoveries in genomics. J. Doudna, "Combining AI and CRISPR Will Be Transformational," *WIRED*, November 26, 2024, https://www.wired.com/story/combining-ai-and-crispr-will-be-transformational

54   In December 2023, the FDA approved Casgevy, the first genome-editing therapy for sickle cell anemia based on CRISPR-Cas9 technology. MIT News, "From CRISPR Breakthroughs, Life-Altering Therapies at MIT," *MIT Spectrum*, Spring 2024, https://betterworld.mit.edu/spectrum/issues/2024-spring/from-crispr-breakthroughs-life-altering-therapies-at-mit/.

55   Researchers have developed a system for bidirectional epigenetic editing, which applies activating and repressive CRISPR mechanisms to gene regulation. H. Zhou et al., "Bidirectional Epigenetic Editing Reveals Hierarchies in Gene Regulation," *Nature Biotechnology* 43 (2025): 355–68, https://www.nature.com/articles/s41587-024-02213-3.

56   Timur Saliev and Prim B. Singh, "From Bench to Bedside: Translating Cellular Rejuvenation Therapies into Clinical Applications." *Cells* 13, no. 24 (2024): 2052, https://doi.org/10.3390/cells13242052.

57   CRISPR technology has opened avenues for correcting genetic disorders in mitochondria. M. Z. Khan et al., "Advances in CRISPR-Cas Technology and Its Applications," *Molecular Therapy—Nucleic Acids*, 2024, https://www.sciencedirect.com/science/article/pii/S2162253123001234.

58   Tech titans are now underwriting longevity research at scale. Altos Labs launched in 2022 with about $3 billion in backing that press reports credit in part to Jeff Bezos. Ian Sample, "If They Could Turn Back Time: How Tech Billionaires Are Trying to Reverse the Ageing Process," *The Guardian*, February 17, 2022, https://www.theguardian.com/science/2022/feb/17/if-they-could-turn-back-time-how-tech-billionaires-are-trying-to-reverse-the-ageing-process. Peter Thiel pledged $3.5 million to the Methuselah Founda-

tion's antiaging prize in 2006. Brandon Keim, "PayPal Founder Wants to Live Forever," *WIRED*, September 18, 2006, https://www.wired.com/2006/09/paypal-founder-wants-to-live-forever/. Google founders set up Calico and—with AbbVie—committed up to $1.5 billion for age-related drug discovery. Ransdell Pierson, "Google's Calico, AbbVie Forge Deal Against Diseases of Aging," *Reuters*, September 3, 2014, https://www.reuters.com/article/us-health-calico-abbvie-idUSKBN0GY24H20140903. And Sam Altman personally invested $180 million in Retro Biosciences in 2022. Devin Coldewey, "Sam Altman Invested $180M in Longevity Startup Retro Biosciences," *TechCrunch*, March 10, 2023, https://techcrunch.com/2023/03/10/sam-altman-invested-180m-in-longevity-startup-retro-biosciences/.

59   A. D. N. J. de Grey, B. N. Ames, J. K. Andersen, A. Bartke, J. Campisi, C. B. Heward et al., "Time to Talk SENS: Critiquing the Immutability of Human Aging, *Annals of the New York Academy of Sciences* 959, no. 1 (2002): 452–62.

60   William Butler Yeats, "Sailing to Byzantium," *The Tower* (Macmillan, 1928), lines 17–24.

61   Estimates suggest a global population of 500,000 to 1 million furries and 2 to 5 million cosplayers. See C. N. Plante et al., "Furscience: A Decade of Psychological Research on the Furry Fandom," *International Anthropomorphic Research Project*, 2020, www.furscience.com, for data on furries, and ZipDo, "Cosplay Industry Statistics 2024," *ZipDo Education Reports*, www.zipdo.co, for cosplay trends showing millions globally.

62   See, for example, Matt Johnson, "The Neuroscience of Mind Uploading and the Psychology of the Digital Afterlife," *Neuroscience of Human Nature* blog, May 9, 2024, https://www.neuroscienceof.com/human-nature-blog/neuroscience-mind-uploading-psychology-digital-afterlife. The idea of transcending our bodies by moving into an inorganic substrate while preserving our consciousness, memory, and coherent sense of self is what transhumanists (particularly those engaged in cryonics) mean when they speak of "uploading." In a later endnote in this section,

we discuss the implications of the likely near-term arrival of weaker versions of that "strong" upload—digital versions of ourselves who behave like we do, "think" like we do, and are largely indistinguishable from us to others, but are not us.

63   Ray Kurzweil, a pioneering inventor and longtime advocate of the concept of exponential technological growth, has predicted the emergence of human-level AI for decades—and with remarkable accuracy. He envisions mind uploading and the merging of biological and machine intelligence not only as plausible, but inevitable, and believes it will happen within his lifetime. In *The Singularity Is Nearer* (2024), a major update of his earlier work, he lays out a detailed timeline toward that future. A co-founder of Singularity University and senior figure at Google, Kurzweil received the National Medal of Technology and Innovation and developed the first reading machines for the blind. Ray Kurzweil, *The Singularity Is Nearer: When We Merge with AI* (Viking, 2024).

64   Robert Doering and Yoshio Nishi, eds., *Handbook of Semiconductor Manufacturing Technology*, 2nd ed. (CRC Press/Taylor & Francis Group, 2007).

65   Preppers assume that order will eventually be restored. They don't try to truly be independent, or they'd need vastly more than go-bags and survival stockpiles. They'd have to prepare to weave their own clothes, grow their own food, mine their own metals, make their own tools, fashion their own weapons, and so much more—all from scratch.

66   *Uploading*, as described in the endnote on "mind uploading" earlier in this section presumes that our consciousness, memory, and coherent sense of self would be preserved, and that we'd know a digital existence akin to immortality. This "strong" upload is a heavy technical lift that may or may not be possible, but even in the best case, it is not close at hand. There is, however, a weaker version of such "strong" upload possibilities. We might create a digital version of ourselves that behaves like we do, "thinks" like we do, and is largely indistinguishable from us to others. If we are alive when that happens, then that avatar is the

AI double we've discussed throughout this book, because we too are present. But if we create that very same avatar at or after our death—so that we are no longer present as our bodily selves—then isn't this AI persona in a sense our "upload," rather than a mere avatar? Maybe not, if it is just mimicking us and knows that. But what if it is also deceiving itself because it has a simulated inner voice whispering that it is our heir, like the sci-fi replicons in *Blade Runner* who think they are human until they learn that their memories are all simulated? And if external observers—our family and friends perhaps—can't tell the difference, and the "upload" can't tell the difference, then in a practical sense, perhaps it really is our "upload" . . . whether or not there is actual continuity of consciousness with us. After all, our own memories are not verifiable reality—some are constructed, some are from our dreams, some are cognitive glitches and distortions. Such "weak" upload seems inevitable in the very near term, even though "strong" uploading is uncertain. I refer to both merely as "uploads," because there will not be a duality here, but a full spectrum between the two—and one that is highly ambiguous. The irony is that the experience of both the upload and its human friends will be the same whether the uploads are "weak" or "strong," so societal impacts will be quite powerful, even for weak uploads.

67   K. Atta et al., "Delusional Misidentification Syndromes: Separate Disorders or Unusual Presentations of Existing DSM-IV Categories?" *Psychiatry (Edgmont)* 3, no. 9 (2006): 56–61.

68   Susan Schneider has explored the philosophical gap between simulating thought and experiencing it, warning that behavioral fidelity may obscure an absence of subjectivity. But whether post-biological minds—however convincing—are truly "us" is similar to trying to determine whether we are real or simulations. How could we know? Susan Schneider, *Artificial You: AI and the Future of Your Mind* (Princeton University Press, 2019).

69   Alcor Life Extension Foundation, "Alcor Membership Statistics," *Alcor*, November 30, 2024, https://www.alcor.org/library/

alcor-membership-statistics; "Membership Statistics," Cryonics Institute, 2024, https://cryonics.org/member-statistics, accessed February 6, 2025.

70   Benjamin Franklin, "Letter to Jacques Dubourg, April 23, 1773," in *The Works of Benjamin Franklin*, vol. 6, ed. John Bigelow (G.P. Putnam's Sons, 1904), 242–43.

71   As described in the earlier endnote in this section, the first uploads will be the avatar digital twins we use. These increasingly lifelike avatars will function long after the demise of their flesh-and-blood template, just like websites of those who die are sometimes maintained by friends or family for decades. Today grief tech is a just-in-time effort to preserve someone's persona. Tomorrow, it will be displaced by our digital twins, and then perhaps transcend us. Will they one day speak to those attending our own memorial? They already have. Rikki Loftus, "CEO of AI Company Had Avatar of His Late Mother Speak at Her Funeral," *Unilad Tech*, August 9, 2024, https://www.uniladtech.com/news/ai/storyfile-ceo-ai-funeral-stephen-smith-423036-20240809. Initially we will see them as copies, but they are an obvious platform for ever deeper infusions of our essence before and after our death, and eventually we may build them to think they are us. Will it be true? How would we ever know? How would they? It will be much easier to create an uploaded self that thinks it is us than to create one that actually is. The example of the metamorphosis of the caterpillar, which provides the cellular architecture of the butterfly, but little more, is worth reflecting on. Is the butterfly really the caterpillar?

72   The concept of uploading consciousness into a digital realm mirrors religious visions of an afterlife, such as Christianity's heaven, Hinduism's moksha, or Islam's Jannah, where the soul transcends the body for eternal awareness or union with a greater reality. These spiritual afterlives share a focus on persistent consciousness but differ in process and outcome: Christianity emphasizes salvation by grace through faith in Christ, Hinduism envisions cyclical reincarnation and liberation, and Islam foresees resurrection according to one's deeds and Allah's mercy.

In contrast, a cybernetic afterlife lacks divine judgment or a metaphysical ensoulment anchor, being an engineered construct unlike the cosmic moral frameworks of spiritual traditions. Uploading's potential for merging with a collective cyber-consciousness echoes mystical unions but does not speak directly to divine order. This technological transcendence thus resembles spiritual aspirations while diverging in its lack of sacred purpose. See also N. Bostrom, "Are You Living in a Computer Simulation?" *Philosophical Quarterly 53*, no. 211 (2003): 243–55; J. Hick, *Death and Eternal Life* (Harper & Row, 1976).

73   Of course, "afterlives" might be physically embodied as well. Robotics is progressing so rapidly now that it will likely be an adequate locus for such reincarnations. It seems likely, though, that the human form—evolved for biological survival—would seem like a hobbled and unimaginative destination for an advanced cyber being. Good enough for an occasional visit, perhaps, but hardly a long-term destination.

74   Initially, I chose 1,000 years for this window—still just an instant in evolutionary time, yet so long that the technology advances would be achieved if they were feasible. A millennium, though, covers so many human lifespans that it falsely suggests that current AI developments might be of limited relevance to us now. This transformative change is reshaping our lives right now, and it matters to us, even though it could be 50 times slower (slow enough not to impact us forcefully) and still be the same monumental evolutionary transition.

75   See, for example, Bernt Børnich, "Meet NEO, Your Robot Butler in Training | Bernt Børnich | TED," YouTube video, 15:19, *TED*, https://www.youtube.com/watch?v=p3uBMqCPSDk.

76   The progression for AI evolution most commonly described is for humans to develop AGI (general intelligence as good as virtually any human), copy it at scale, and have it push forward to develop ASI that far exceeds all human capacities. This makes sense, as an AGI army of computer scientists working together 24/7 would be a formidable force, but is AGI a critical step? We already can create domain-specific

ASIs that exceed humans in narrow realms (ANI is *artificial narrow intelligence*), like reading radiology images or doing math. If we were to create an ANI team that excelled at computer programming and cognitive modeling, why would their inability to discuss poetry, understand word puzzles, or read human emotion undercut their capacity to build more potent AI that ultimately becomes AGI and ASI? AGI is certainly a step along the path to ASI, but is it an essential element in creating ASI? This question brings up the issue of whether generalized intelligence—the ability to learn from immediate experience rather than the massive, neural-network training behind LLMs—could emerge without having such thinking supplied by bright, creative humans.

77   Here are some arguments against AI consciousness: Roger Penrose, *Shadows of the Mind* (Oxford University Press, 1994), in his chapters discussing Gödelian arguments and non-computability; John Searle, *The Mystery of Consciousness* (New York Review of Books, 1997); Robert J. Marks, *Non-Computable You: What You Do That Artificial Intelligence Never Will* (Discovery Institute, 2022); Robert Epstein, "The Empty Brain," *Aeon*, May 18, 2016. But these are viewed as somewhat fringe now, particularly since whether a perfect simulation of consciousness and emotion would actually "be" conscious is impossible to establish even for us. See Daniel Dennett, *Consciousness Explained* (Little, Brown, 1991) and *The Future of the Brain: Essays by the World's Leading Neuroscientists* (Princeton University Press, 2013).

78   Here are arguments that we are nowhere near radical life extension: S. Jay Olshansky et al., "In Pursuit of the Longevity Dividend: Scientific Goals for an Aging World," *The Gerontologist* 47, no. 3 (2007): 231–39; T. Kirkwood, "Understanding the Odd Science of Aging," *Cell* 120, no. 4 (2005): 437–47. But many are much more optimistic: Matthew D. LaPlante, *Lifespan: Why We Age—and Why We Don't Have To* (Atria Books, 2019); Nir Barzilai, *Age Later: Health Span, Life Span, and the New Science of Longevity* (St. Martin's Press, 2020); Carlos López-Otin, Maria A. Blasco, Linda Partridge, Manuel Serrano, and Guido Kroemer, "Hallmarks of Aging: An Expanding Universe," *Cell* 186, no. 2 (2023): 243–78.

79 If machines can never achieve consciousness, our minds could never be uploaded. Michael Egnor, "Could You Upload Your Mind into a Computer?" *Mind Matters*, May 18, 2021, www.mindmatters.ai; Suzanne Gildert, "Transcend or Transhuman? The Real Barriers to Uploading the Human Mind," *H+ Magazine* (online), 2018.

80 Angela Yang, "Lawsuit Claims Character.AI Is Responsible for Teen's Suicide," *NBC News*, October 23, 2024, https://www.nbcnews.com/tech/characterai-lawsuit-florida-teen-death-rcna176791.

81 Sesame has one of the very best voice engines. See here for a demo: https://www.ai-sesame.com/.

82 American Psychiatric Association, *Diagnostic and Statistical* Manual *of Mental Disorders (DSM-5-TR)*, 5th ed., text revision (American Psychiatric Publishing, 2022).

83 "Schizophrenia," National Institute of Mental Health, NIH, 2023, www.nimh.nih.gov/health/topics/schizophrenia.

84 Graeme J. Taylor, and R. Michael Bagby, "Alexithymia and the Five-Factor Model of Personality," *Comprehensive Psychiatry* 33, no. 3 (1992): 147–51, https://doi.org/10.1016/0010-440X(92)90023-J.

85 David Neary et al., "Frontotemporal Lobar Degeneration: A Consensus on Clinical Diagnostic Criteria," *Neurology* 51, no. 6 (1998): 1546–54, https://doi.org/10.1212/WNL.51.6.1546.

86 Robert D. Hare, *Without Conscience: The Disturbing World of the Psychopaths Among Us* (Guilford Press, 1993).

87 Nitasha Tiku, "The Google Engineer Who Thinks the Company's AI Has Come to Life," *Washington Post*, June 11, 2022.

88 Lucy Mangan, "Eternal You Review–It's Impossible Not to Be Horrified by This AI Quest to Bring the Dead Back to Life," *The Guardian*, October

29, 2024, https://www.theguardian.com/tv-and-radio/2024/oct/29/ storyville-eternal-you-review-film-dead; Dan Milmo, "'I Felt I Was Talking to Him': Are AI Personas of the Dead a Blessing or a Curse?" *The Guardian*, June 14, 2024, https://www.theguardian.com/lifeandstyle/ article/2024/jun/14/i-felt-i-was-talking-to-him-are-ai-personas-of-the-dead-a-blessing-or-a-curse.

89   Distortions in childhood memories are surprisingly common, with studies suggesting that up to 25% of individuals can adopt suggested events as real recollections when prompted. This unreliability, particularly when influenced by others, casts doubt on memory's accuracy in legal contexts like sexual assault cases, where imposed memories can emerge. E. F. Loftus and J. E. Pickrell, "The Formation of False Memories," *Psychiatric Annals* 25, no. 12 (1995): 720–25, https:// journals.healio.com/doi/10.3928/0048-5713-19951201-07.

90   J. Smith, "Behind the Curtain: Deception in The Wizard of Oz," in *Cinematic Illusions*, ed. A. Jones (Academic Press, 2020); V. Fleming, dir., *The Wizard of Oz*, Metro-Goldwyn-Mayer, 1939.

91   Estimates suggest that 10,000 to 30,000 Elvis impersonators are active worldwide as of 2025. The scale of this phenomenon is highlighted by the largest recorded gathering of Elvis impersonators, where 895 performers convened at Harrah's Cherokee Casino Resort on July 12, 2014, earning a Guinness World Record (Guinness World Records, 2014). While these numbers are speculative due to limited primary data, they reflect the enduring cultural impact of Elvis Presley through tribute performances.

92   ABBA Voyage, launched in May 2022 with a $175 million investment, transformed live music through motion capture and AI-driven "ABBAtars," depicting ABBA's 1979 selves, created by Industrial Light & Magic. Hosted in the 3,000-seat ABBA Arena, it sold over 1.5 million tickets by September 2023, generating $150 million in sales. This success points to IMAX-style concert venues—shared arenas hosting rotating virtual acts. By 2030, production costs

could drop to $5–$10 million, using affordable motion capture, AI rendering, and modular LED screens, offering tickets at $30–$50. Financed like multiplexes, these venues would scale globally, amplifying music's communal essence without artist limits, mirroring cinema's accessibility shift. AI-driven interactivity could blur real and virtual performances, redefining concerts as immersive, affordable spectacles. Dylan Smith, "ABBA Voyage Revenue Approached $138 Million in 2023," *Digital Music News*, October 1, 2024, https://www. digitalmusicnews.com/2024/10/01/abba-voyage-revenue-2023/. Wikipedia, "ABBA Voyage," June 1, 2022, https://www.euronews. com/culture/2024/12/10/abba-voyage-gives-uk-economy-huge-boost-contributing-14-billion?utm_source=chatgpt.com https:// en.wikipedia.org/wiki/ABBA_Voyage.

93   See extended endnote on sexbots in chapter 1.

94   In June 2025, Chinese scientists modified stem cell-derived human repair cells to keep the antiaging gene FOXO3 switched on. They call the result "senescence-resistant cells" (SRCs) because the cells don't wear out as quickly. Four monthly IV infusions of SRCs were given to middle-aged monkeys. After treatment, the animals had 40%-60% fewer "senescent" (worn-out) cells in blood, liver, and muscle. They also remembered tasks better, gripped harder, handled sugar more smoothly, and even showed signs of hormone recovery. DNA "aging clocks" suggested their biological age rolled back by more than six years. The authors propose that, if safety holds up, getting an occasional SRC "top-up" could one day serve as a practical antiaging therapy for people. Jinghui Lei et al., "Senescence-Resistant Human Mesenchymal Progenitor Cells Counter Aging in Primates," *Cell* 188, no. 18 (2025): P5039–61, https://doi. org/10.1016/j.cell.2025.05.021.

95   J. Jumper, R. Evans, A. Pritzel et al., "Highly Accurate Protein Structure Prediction with AlphaFold," *Nature* 596, no. 7873 (2021): 583–89, https://www.nature.com/articles/s41586-021-03819-2.

96   Jeffrey Kluger, "The Return of the Dire Wolf," *Time*, April 7, 2025, https://time.com/7274542/colossal-dire-wolf/.

Chapter 8: The Mother of All Paradigm Shifts

1   Gould analyzes the transition from single-celled eukaryotes to multicellular metazoans as a pivotal evolutionary leap, concluding that such shifts redefine life's complexity, and assumes rapid diversification follows structural change. Stephen J. Gould, *Wonderful Life: The Burgess Shale and the Nature of History* (W. W. Norton, 1989).

2   Kuhn introduces the concept of paradigm shifts, concluding that scientific progress often hinges on revolutionary breaks rather than gradual gains, and assumes shared assumptions shape entire fields. T. S. Kuhn, *The Structure of Scientific Revolutions* (University of Chicago Press, 1962).

3   Popper argues scientific progress comes through incremental testing and refutation, while Shapin suggests change is often subtler than dramatic leaps, both concluding that Kuhn's model overstates ruptures, and assuming steadier paths dominate science. K. Popper, *Conjectures and Refutations: The Growth of Scientific Knowledge* (Routledge, 1963); S. Shapin, *The Scientific Revolution* (University of Chicago Press, 1996).

4   Kuhn's earlier book traced Copernicus's heliocentric model as a radical shift in cosmology, and his later book tied Einstein's Relativity to similar upheavals, concluding that such changes redefine truth, and assuming resistance delays their acceptance. T. S. Kuhn, *The Copernican Revolution: Planetary Astronomy in the Development of Western Thought* (Harvard University Press, 1957); Kuhn, *Structure of Scientific Revolutions*.

5   Kuhn outlines how paradigms shift through anomalies, crises, and new frameworks, concluding that this cycle drives major scientific advances,

and assumes old and new paradigms rarely coexist smoothly. Kuhn, *Structure of Scientific Revolutions*.

6   Hobsbawm examines the Industrial Revolution's transformation from craft to factory economies, concluding that technology amplified human labor's role, and assumes scarcity persisted as a core driver. E. J. Hobsbawm, *Industry and Empire: From 1750 to the Present Day* (Penguin Books, 1968).

7   Kuhn, *Copernican Revolution*.

8   These five key "truths" underlie the dominant paradigm of society today, and they are collapsing: (1) *Human appetites are limitless* so we will always be buying more goods and services and experiences. But human time and attention are limited, and we're increasingly overwhelmed by information. The most consumptive societies like the United States seem to measure less on scales of thriving than those that are modern but less consumptive. (2) *There will always be scarcity*, and it will get worse as we deplete resources. But as energy gets cheaper and we have widespread AI, abundance will become the dominant story. (3) *Human intelligence is unachievable by AI any time soon.* But advanced AI systems now outperform us in data analysis, artistic creation, and even emotional understanding. For Gen AI, raised with AI tools from infancy, this will be obvious. (4) *Humans are flesh and blood, and that will always be the case.* But integration with AI, from neural implants to enhanced senses, will almost certainly lead us to hybrid forms of existence beyond the current functional cyborgian unions already occurring. (5) *We are the masters and shapers of our technology.* But AI's autonomy, evident in its independent problem-solving and exploration of new domains, suggests that it may soon be the controlling partner within our planetary union.

9   USDA data show ~70% of US jobs were agricultural in 1920, dropping sharply by 1940, while BLS notes ~70% of 2023 jobs are

knowledge-based, concluding that labor shifts redefine economies, and assuming new roles emerge with change. U.S. Department of Agriculture, *Census of Agriculture: 1920* (U.S. Government Printing Office, 1920); Bureau of Labor Statistics, "Employment by Industry, 2023 Annual Averages" (U.S. Department of Labor, 2023).

10   Kissinger et al. examine AI's potential to reshape society and alter power dynamics within existing frameworks. H. A. Kissinger, E. Schmidt, and D. Huttenlocher, *The Age of AI: And Our Human Future* (Little, Brown, 2021).

11   Kuhn details how Ptolemy's geocentric model failed under mounting astronomical anomalies, concluding that Copernicus's heliocentrism marked a paradigm shift, and assumes such shifts require bold rethinking. Kuhn, *Copernican Revolution*.

12   Westfall chronicles the decades-long debate over Copernicus's heliocentric model, concluding that Galileo's advocacy by 1610 solidified its impact, and assumes resistance stemmed from entrenched beliefs. R. S. Westfall, *Never at Rest: A Biography of Isaac Newton* (Cambridge University Press, 1980).

13   Lee and Qiufan envision AI transforming education with tools like tutors, concluding that learning will become more personalized, and assume schools will integrate AI without collapsing. K.-F. Lee and C. Qiufan, *AI 2041: Ten Visions for Our Future* (Crown Currency, 2021).

14   BLS data confirm ~70% of US jobs are now service-based, reflecting modern economies, while Hobsbawm notes the historical rise of factory labor; together concluding that economic structures evolve with technology, and assuming human work remains central. Bureau of Labor Statistics, "Employment by Industry, 2023 Annual Averages" (U.S. Department of Labor, 2023); Hobsbawm, *Industry and Empire: From 1750 to the Present Day* (Penguin Books, 1968).

15   Davidow and Malone highlight AI's ability to surpass human skills like coding, concluding that automation will disrupt traditional jobs and will force societies to adapt to new labor models. W. H. Davidow and M. S. Malone, *The Autonomous Revolution: Reclaiming the Future We've Sold to Machines* (Berrett-Koehler, 2020).

16   A landmark account of how Deep Blue beat Garry Kasparov, demonstrating early AI's dominance over symbolic, high-level intellectual domains like chess. Murray Campbell, A. Joseph Hoane Jr., and Feng-hsiung Hsu, "Deep Blue," *Artificial Intelligence* 134, nos. 1–2 (2002): 57–83, https://doi.org/10.1016/S0004-3702(01)00129-1.

17   Einstein's work establishes relativity's new framework of space-time, while Kuhn ties it to incommensurable shifts, concluding that such changes redefine scientific truth, and assuming old metrics become obsolete. A. Einstein, *Relativity: The Special and General Theory* (Methuen, 1916); Kuhn, *Structure of Scientific Revolutions*.

18   Rhodes recounts Hiroshima's 1945 detonation, concluding that atomic power introduced incommensurable scales of destruction and that humanity struggled to grasp its new role. R. Rhodes, *The Making of the Atomic Bomb* (Simon & Schuster, 1986).

19   Kissinger et al. note AI's shift in social dynamics, concluding that trust and interactions will evolve with technology, and assume human relationships will adapt to mediation. Kissinger et al., *Age of AI*.

20   AI systems are now surpassing physician-AI teams in radiology and pathology. E. J. Topol, "High-Performance Medicine: The Convergence of Human and Artificial Intelligence," *Nature Medicine* 25, no. 1 (2019): 44–56. This will dramatically change the role of physicians in ways that can be generalized to other knowledge-based professions.

21   Lee, Tegmark, Davidow, Malone, Crawford, Kissinger, Schmidt, and Huttenlocher collectively explore AI's impact on work, learning, and society, concluding that it will reshape human systems. They, however,

assume adaptation will occur within familiar human roles. Lee and Qiufan, *AI 2041*; M. Tegmark, *Life 3.0: Being Human in the Age of Artificial Intelligence* (Knopf, 2017); Davidow and Malone, *Autonomous Revolution*; Kate Crawford, *The Atlas of AI: Power, Politics, and the Planetary Costs of Artificial Intelligence* (Yale University Press, 2021); Kissinger et al., *Age of AI*.

22   Tegmark frames AI as a potential new phase of existence, concluding that it could redefine life's trajectory, and assumes humans must guide its evolution thoughtfully. Tegmark, *Life 3.0*.

23   Gould explains how Darwin's natural selection humbled humanity by placing us among other species, concluding that it reshaped our self-understanding, and assumes such shifts alter cultural truths. S. J. Gould, *The Structure of Evolutionary Theory* (Harvard University Press, 2002).

24   Lee, Quifan, Kissinger, Schmidt, and Huttenlocher project AI's near-term paths, concluding that it will transform daily life and assume humans will navigate these changes with current frameworks. Lee and Qiufan, *AI 2041*; Kissinger et al., *Age of AI*.

25   V. Vinge, *Marooned in Realtime* (Bluejay Books, 1986); V. Vinge, "The Coming Technological Singularity: How to Survive in the Post-Human Era," in *Vision-21: Interdisciplinary Science and Engineering in the Era of Cyberspace*, ed. G. A. Landis, NASA Conference Publication 10129 (NASA Lewis Research Center, 1993), 11–22.

26   For a rich exploration of this, see John Smart, "Intro to the Developmental Singularity Hypothesis (DSH)," *AccelerationWatch.com*, first presented at Foresight Vision Weekend, Palo Alto, May 20, 2000, https://www.accelerationwatch.com/developmentalsinghypothesis.html

27   It is often asserted that the distribution of wealth in Western societies is dramatically and increasingly skewed toward the wealthy few—the very top percentiles of the population. See, for example, Piketty's classic: T.

Piketty, *Capital in the Twenty-First Century*, trans. A. Goldhammer (Belknap Press of Harvard University Press, 2014). But when welfare and retirement accounts and home ownership are included in these calculations, this skew toward massive inequality fades, and it is clear that societal wealth is increasingly widely distributed. This is discussed at length by Waldenstrom in his recent book. D. Waldenström, *Richer and More Equal: A New History of Wealth in the West* (Polity Press, 2024), and sketched in his article in *Foreign Affairs*. D. Waldenström, "The Inequality Myth: Western Societies Are Growing More Equal, Not Less," *Foreign Affairs*, May 19, 2025, https://www.foreignaffairs.com/united-states/inequality-myth.

Chapter 9: A New Order Emerges

1   For a detailed analysis of the prospects for a rapid, disruptive transition to abundance driven by AI replacement of virtually all human labor, see E. Erdil and T. Besiroglu, "Explosive Growth from AI Automation: A Review of the Arguments," arXiv preprint, 2023. In 2025, Besiroglu founded a company whose goal is to replace all human labor. See https://techcrunch.com/2025/04/19/famed-ai-researcher-launches-controversial-startup-to-replace-all-human-workers-everywhere/.

2   N. E. Borlaug, "The Green Revolution Revisited and the Road Ahead," *Annals of the New York Academy of Sciences* 976, no. 1 (2002): 297–300. In 1840, 70% of the US labor force was associated with farming. Today the figure is 1.6%. National Agricultural Statistics Service, *2022 Census of Agriculture* (U.S. Department of Agriculture, 2023), https://www.nass.usda.gov/Publications/AgCensus/2022/.

3   See P. H. Diamandis and S. Kotler, *The Future Is Faster Than You Think* (Simon & Schuster, 2020); E. Brynjolfsson and A. McAfee, *The Second Machine Age* (W. W. Norton, 2014).

4   Walter Scheidel, "Human Mobility in Roman Italy, II: The Slave Population," *Journal of Roman Studies* 95 (2005): 64–79, https://doi.org/10.3815/000000005784016270; M. H. Hansen, *The Shotgun Method: The Demography of the Ancient Greek City-State Culture* (University of Missouri Press, 2006).

5   D. J. Thompson, "Slavery in the Hellenistic World," in *The Cambridge World History of Slavery: Volume 1*, ed. K. Bradley and P. Cartledge (Cambridge University Press, 2011), 194–213; R. Miles, *Carthage Must Be Destroyed: The Rise and Fall of an Ancient Civilization* (Penguin Books, 2011).

6   See Y. N. Harari, *Homo Deus: A Brief History of Tomorrow* (Harper, 2015); E. Klein, *Abundance* (Penguin Press, 2024); Brynjolfsson and McAfee, *Second Machine Age*; S. Zuboff, *The Age of Surveillance Capitalism* (PublicAffairs, 2019).

7   In the United States, a median household earns about $80,000 yearly, and for only about $2,000 a year in subscription fees, we could have all these digital wonders. U.S. Census Bureau, *Income and Poverty in the United States: 2023* (U.S. Government Publishing Office, 2023).

8   Travel from New York to Mumbai in 1800 cost about $8,000 in 2025 dollars for first class, spanning roughly 125 days on sailing ships around the Cape of Good Hope. First-class cabins provided bunks and simple meals but lacked sanitation, offering grim, monotonous, dangerous conditions by today's standards. In 1900, first class cost $17,650 and took 23 days, using ocean liners to London, trains to Brindisi, and steamers via the Suez Canal, with elegant dining and cabins yet poor hygiene and a tedious duration. Today's economy class flight, at $400, takes 15.5 hours nonstop, delivering seats, meals, screens, and clean restrooms—comforts surpassing, in speed and cleanliness, the 1800's squalor and 1900's dated luxury. Modern economy class outshines historical first class, costing 20 times less and at a pace 200 times faster than 1800, and 44 times cheaper and 36 times quicker than 1900. These costs and conditions are based on maritime and rail records, including Cunard and P&O Line fares, adjusted to 2025 dollars via CPI data (U.S. Bureau of Labor

Statistics, 2025) and Grok 3. Note that neither a Rockefeller nor a king could improve the experience described here for 1800 or 1900. The technology just wasn't there.

9   Basic food this cheap isn't here yet, but AI makes it possible. Vertical farms cut costs; leafy greens are $10/kg now, heading to $2/kg by 2035 with AI optimizing water and nutrients. Fortified staples like rice add vitamins for $0.10/kg, and AI logistics slash waste. Lab-grown proteins could hit $0.20/meal soon. It wouldn't be a feast, but it would at least provide the bare-bones nutrition of grains and proteins. Granted, global politics is a significant obstacle, but solving the worldwide hunger problem is feasible. N. Bowles, "AI-Driven Vertical Farming: Cost Reductions and Scalability," *Journal of Agricultural Innovation* 12, no. 3 (2024): 88–97; Z. A. Bhutta et al., "Biofortification of Staple Crops: A Sustainable Solution to Micronutrient Deficiencies," *The Lancet Global Health* 8, no. 4 (2020): e524–32.

10   A "safe, clean bunk for everyone" would mean a minimal safety net, not necessarily a home. 3D-printed units cost $10,000 now, maybe $1,000 by 2035 with AI cutting labor. Think capsule hotels—50 square feet, bed, locker—scaled globally. AI monitoring, like Shenzhen's dorms, adds safety for $0.10/day, catching fires or fights. This assumes need, not universal use, and if 20 million were housed in India recently, AI could push this further, fast. Apis Cor, "3D-Printed Housing: Cost and Scalability Projections," *Construction Technology Review* 9, no. 2 (2024): 34–41; L. Zhang et al., "AI-Driven Security in Urban Dormitories: Shenzhen Case Study," *Smart Cities Journal* 6, no. 4 (2023): 112–20.

11   World Bank, *Poverty and Shared Prosperity 2022* (World Bank Group, 2022), https://openknowledge.worldbank.org/server/api/core/bitstreams/ b96b361a-a806-5567-8e8a-b14392e11fa0/content

12   Capitalism's success in elevating global living standards is vividly shown by Hans Rosling, who claimed extreme poverty plummeted from 85% in 1800 to under 10% by 2017, with billions lifted into a "big middle" via income growth and health gains, depicted in dynamic

graphs. H. Rosling, O. Rosling, and A. Rosling Rönnlund, *Factfulness: Ten Reasons We're Wrong about the World—And Why Things Are Better Than You Think* (Flatiron Books, 2018); H. Rosling, "Don't Panic—How to End Poverty in 15 Years," TV documentary, *BBC Two*, September 22, 2015. Be sure to watch this brilliant short video of his: Hans Rosling, "200 Countries, 200 Years, 4 Minutes—The Joy of Stats," YouTube video, 4:48, posted November 26, 2010, https://www.youtube.com/watch?v=jbkSRLYSojo.

13   S. Garasky, K. Mbwana, A. Romualdo, A. Tenaglio, and M. Roy, *Foods Typically Purchased by Supplemental Nutrition Assistance Program (SNAP) Households*, (U.S. Department of Agriculture, Food and Nutrition Service, 2016), https://fns-prod.azureedge.us/sites/default/files/ops/SNAPFoodsTypicallyPurchased-Summary.pdf.

14   There are perhaps 8 million people in the United States without phone service due to poverty. E. A. Vogels, "A Digital Divide Persists Even as Americans with Lower Incomes Make Gains in Tech Adoption," Pew Research Center, June 22, 2021, https://www.pewresearch.org/short-reads/2021/06/22/digital-divide-persists-even-as-americans-with-lower-incomes-make-gains-in-tech-adoption/.) At $300/phone and a subscription plan of $40/month, that would be $6.4 billion the first year and less in later years. Plus it would include a digital camera, a calculator, a navigation aid, etc. And for another few hundred annually, LLM access to most human knowledge, instantaneous translation, medical advice, coaching (including for mental health), and much more. The biggest missing pieces are transportation, shelter, and food, but paths toward those aren't so far away with ride-sharing, 3D printing technology, and robotic farming around the corner.

15   For example, see this paper on its use in voting. Y. Miao, "Secure and Privacy-Preserving Voting System Using Zero-Knowledge Proofs," Proceedings of the International Conference on Software Engineering and Machine Learning, 2023, https://www.researchgate.net/publication/372823212_Secure_and_Privacy-Preserving_Voting_System_Using_Zero-Knowledge_Proofs.

16   Modest abundance in the United States requires around $120,000–$145,000 posttax ($170,000–$205,000 pretax at a 30% tax rate) for a family of four. Estimated annual expenses include housing ($18,000–$42,000), food ($18,000), healthcare ($24,000), transportation ($12,000), education ($12,000), discretionary spending ($36,000). At a conservative 4% withdrawal rate, this requires liquid assets totaling about $4.2–$5.2 million excluding a primary residence ($750,000). Luxurious abundance demands perhaps $900,000–$1 million posttax annually ($1.8 million–$2 million pretax) to cover housing ($125,000–$250,000), food ($75,000), healthcare ($60,000), transportation ($120,000), education ($120,000), discretionary spending ($400,000). This requires liquid assets of about $36 million–$40 million at a 5% withdrawal rate, excluding primary residences ($5 million). US wealth data indicate approximately 1.5% of households (2 million total) have enough assets for modest abundance and 0.15% (200,000 households) have enough for luxurious abundance. The total number of households globally with this wealth and affluence are a little more than twice the US figures. All figures were assembled using ChatGPT-4o to access diverse sources, including Federal Reserve Board, *Survey of Consumer Finances (SCF), 2022* (Board of Governors of the Federal Reserve System, 2023), https://www.federalreserve.gov/econres/scfindex.htm; Credit Suisse, *Global Wealth Report 2022* (Credit Suisse Research Institute, 2022), https://www.credit-suisse.com/about-us/en/reports-research/global-wealth-report.html.

17   Observations drawn from the Federal Reserve Board, *Survey of Consumer Finances (SCF), 2022*; Credit Suisse, *Global Wealth Report 2022*.

18   W. Gibson, "The Science in Science Fiction," radio broadcast interview, *Talk of the Nation*, National Public Radio, November 30, 1999.

19   Some argue that abundance, by making so much available, only sharpens our hunger for what we don't have—status, exclusivity, the next new thing. They see it breeding envy or a restless chase for meaning in all the wrong places. But this assumes we're trapped by desire. If abundance is a lens, not a tally, the problem isn't plenty itself. Harari, *Homo Deus*.

20   Tesla predicts that a consumer version of the Optimus humanoid robot will ship in 2026, cost about $20,000, perform 1,000 different household tasks, and hit annual production rates over a million. "Tesla Optimus Update—Price, Release Date & 1,000-Task Demo," YouTube video, posted by "Dr. Know-it-All Knows It All," April 15, 2025, https://www.youtube.com/watch?v=suV7GE4nvNs. This seems more hype than possibility, but is it likely in 5 years? Or 10? The cognitive power required to work in a single home would be less than autonomous driving today in unpredictable, real-world environments with countless vehicles and pedestrians, and the consequences of mistakes are far less. Production would be much cheaper than a car too, so the price does not sound unreasonable. The appetite for this would be massive. And soon, multitudes would enjoy having personal servants to cook, do laundry, clean, fix breakfast, lay out clothes, tell them their schedules, and more. See *The Crown*, created by Peter Morgan (Netflix, 2016–2023).

21   Critics warn that technology, far from solving abundance's voids, could deepen them—luring us with shallow pleasures or binding us through surveillance. They fear a world where AI amplifies distraction or turns us into data that drive profit. But this assumes tech stays tethered to today's motives. In a true abundance economy, with AI companions and protectors, the power of advertising-manufactured desire might weaken. We might test diverse AI guides *in silico* to ensure they liberate rather than control us—the same qualities we will seek in our children's AI tutors. A. Huxley, *Brave New World* (Chatto & Windus, 1932); Zuboff, *Age of Surveillance Capitalism*.

22   Centers for Disease Control and Prevention, "Obesity Prevalence in the United States, 2017–2023," National Center for Health Statistics, 2023.

23   Michelle Faverio, Monica Anderson, and Eugenie Park, "Teens, Social Media and Mental Health," Pew Research Center, April 22, 2025, https://www.pewresearch.org/internet/2025/04/22/teens-social-media-and-mental-health/.

24   Humanity has long wrestled with the concept of material abundance, noting its promise and its many traps. Many have argued it could fuel endless desire rather than satisfaction, widen gaps between haves and have-nots, erode the drive to strive, or worse, pull us into distraction and waste, while bringing up issues of inequality, sustainability, consumerism, corporate control, and social division. Yet abundance is not just about having more; it's about having enough. Monks and ascetics find plenty in simplicity, thriving not on goods but on existence. See P. H. Diamandis and S. Kotler, *Abundance: The Future Is Better Than You Think* (Free Press, 2012).

25   Some insist that meaning comes from human struggle—through community, sacrifice, and facing life's challenges. Others see machines as cold and lacking the depth of a friend or sage. These concerns reflect doubts about AI's ability to think or feel, but they are fading as systems progress. And spiritual epiphanies can happen in an instant, so engaging with an AI spiritual guide need not be long term. AI could scale what humans offer without human limits. C. Taylor, *A Secular Age* (Harvard University Press, 2007); V. E. Frankl, *Man's Search for Meaning* (Beacon Press, 1946).

26   The risk of misuse is real, but this challenge seems surmountable. For perspective, see Brynjolfsson and McAfee, *Second Machine Age*; Zuboff, *Age of Surveillance Capitalism*.

27   Woebot Health, "Efficacy of AI-Based Mental Health Interventions: A Meta-Analysis," *Nature Digital Medicine* 7, no. 1 (2024): 45–53.

28   The problem comes when we think that more will make us whole—a trap that leaves us empty. The issue isn't abundance itself, but how we deal with it. V. E. Frankl, *Man's Search for Meaning* (Beacon Press, 1946). M. E. P. Seligman, *Flourish: A Visionary New Understanding of Happiness and Well-Being* (Simon & Schuster, 2021).

29   Note that there is overlap in these categories. A skilled heart surgeon falls in both wave 2 and 3, having both deep knowledge and one-off manual

dexterity challenges, while a radiologist or psychiatrist is squarely in wave 2 and more vulnerable.

30   Erik Brynjolfsson, Bharat Chandar, and Ruyu Chen, "Canaries in the Coal Mine? Six Facts about the Recent Employment Effects of Artificial Intelligence," Stanford Digital Economy Lab, August 26, 2025, https://digitaleconomy.stanford.edu/wp-content/uploads/2025/08/ Canaries_BrynjolfssonChandarChen.pdf.

31   "Tesla Optimus Update—Price, Release Date & 1,000-Task Demo," YouTube video, posted by "Dr. Know-it-All Knows It All," April 15, 2025, https://www.youtube.com/watch?v=suV7GE4nvNs. (See the extended endnote about this in the Boundaries of Plenty section.) Chinese robotics firm Unitree unveiled a nimble humanoid "intelligent companion" in July 2025 at the astonishing price of $5,900. Unitree Robotics, "Unitree R1 Intelligent Companion Price from $5900," YouTube video, posted July 24, 2025, https://www.youtube.com/ watch?v=v1Q4Su54iho. The consumer version of the Optimus humanoid robot is projected to ship in 2026, cost about $20,000, perform 1,000 different household tasks, and hit annual production rates over a million. "Tesla Optimus Update—Price, Release Date & 1,000-Task Demo," YouTube video, posted by "Dr. Know-it-All Knows It All," April 15, 2025, https://www.youtube.com/watch?v=suV7GE4nvNs. The timeline for the Optimus humanoid robot by Tesla may be a bit aggressive, but the cognitive power required to work in a single home would be less than today's autonomous driving in unpredictable, real-world environments with countless vehicles and pedestrians, and the consequences of mistakes are far less. Production would be much cheaper than a car, so the price sounds reasonable. The appetite for this would be massive. And soon, multitudes would enjoy having personal servants to cook, do laundry, clean, fix breakfast, lay out clothes, tell them their schedules, and more. See *The Crown* (created by Peter Morgan, Netflix, 2016–2023).

32   This exploration of the collapse of our dominant propositional-knowledge framework and its existential consequences is excellent, though it

fails to fully appreciate the extent of what is now underway. Nicolas Michaelsen, "Death of a Knowledge System," *Ecologies of Wisdom*, Substack, May 6, 2025, https://ecologiesofwisdom.substack.com/p/death-of-a-knowledge-system.

33   In March 2025, Open AI announced that it planned to offer PhD-level agents at $20,000 per month. If this happens, it would make it possible to bring together talented, diverse, complex teams quickly to aggressively explore and plan new projects. See https://techstartups.com/2025/03/05/openai-is-planning-to-charge-20000-a-month-for-phd-level-ai-agents/.

34   A recent MIT Media Lab preprint investigates how extended use of ChatGPT can diminish neural engagement and memory retention. Nataliya Kosmyna et al., "Your Brain on ChatGPT: Accumulation of Cognitive Debt When Using an AI Assistant for Essay Writing Task," *MIT Media Lab*, June 2025, https://doi.org/10.48550/arXiv.2506.08872.

35   Vincent Chin, Noelle Samia, Roman Marchant et al., "A Case Study in Model Failure? COVID-19 Daily Deaths and ICU Bed Utilisation Predictions in New York State," arXiv, June 26, 2020, https://doi.org/10.48550/arXiv.2006.15997.

36   Paul Ehrlich, *The Population Bomb* (Ballantine Books, 1968).

37   Commission on the Intelligence Capabilities of the United States Regarding Weapons of Mass Destruction, "Report to the President," US Government Printing Office, March 31, 2005, https://policy.defense.gov/portals/11/Documents/hdasa/references/GPO-WMD.pdf.

38   Anjan V. Thakor, "The Financial Crisis of 2007–2009: Why Did It Happen and What Did We Learn?" *Review of Corporate Finance Studies* 4, no. 2 (2015): 155–205, https://doi.org/10.1093/rcfs/cfv001.

39   "Israel Showed That Seizing Air Superiority Isn't Gone from Modern
      Warfare, but Iran Isn't China or Russia," *Business Insider*, June 2025, https://
      www.businessinsider.com/israel-shows-air-superiority-possible-war-
      still-unlikely-against-china-2025-6.

40   Tetlock's work shows that experts often perform worse than simple
      algorithms, highlighting the overconfidence of the intellectual
      elite. Philip E. Tetlock, *Expert Political Judgment: How Good Is It? How
      Can We Know?* (Princeton University Press, 2017), https://doi.
      org/10.1515/9781400830312.

Chapter 10: A Cosmic Lens

1   Brian Swimme offers a fine poetic account of the universe as a dynamic,
     evolving entity in which human self-awareness emerges as part of
     a grand evolutionary arc. His concept of "cosmogenesis" aligns with
     the birth of Metahumanity in an unfolding universe that is discovering
     itself. Brian Swimme and Thomas Berry, *The Universe Story: From the
     Primordial Flaring Forth to the Ecozoic Era—A Celebration of the Unfolding of
     the Cosmos* (Harper, 1992).

2   My own avatar, started with a few PDFs, will vastly improve as I feed
     it more of my conversations. What costs $25,000 to do well today will
     soon cost a tiny fraction of that and will become ordinary, no more
     remarkable than having a disembodied voice in our car to prompt us
     as we drive . . . or an AI coach to encourage us as we exercise.

3   Clément Vidal and I explored this directly in the conversation we had
     with ChatGPT mentioned in the introduction.

4   We see difficulties with current models that haven't even achieved
     general intelligence, so what hope do we have for ASI? Beatrice Nolan,
     "Leading AI Models Show Up to 96% Blackmail Rate When Their

Goals or Existence Is Threatened, Anthropic Study Says," *Fortune,* June 23, 2025, https://archive.is/pZH84.

5   D. M. Raup, *Extinction: Bad Genes or Bad Luck?* (W.W. Norton, 1991); M. J. Benton, *The Dinosaurs Rediscovered* (Thames & Hudson, 2019). Mammalian species survive about 1 million years on average (R. L. Smith, "The Evolution of Vertebrate Diversity," *People & the Planet* 4, no. 3 [1995]: 6-9), though some like moles may have survived as long as 10 million years, and particularly resilient species like cockroaches, horseshoe crabs, and silverfish have survived 50, 150, and 300 million years respectively. David Grimaldi and Michael S. Engel, *The Evolution of Insects* (Cambridge University Press, 2005); C. N. Shuster Jr., R. B. Barlow, and H. J. Brockmann, eds., *The American Horseshoe Crab* (Harvard University Press, 2003).

6   The median survival of species in the Homo genus has been about 500,000 years. D. Reich et al., "Genetic History of an Archaic Hominin Group from Denisova Cave in Siberia," *Nature* 468, no. 7327 (2010): 1053–60; R. E. Green et al., "A Draft Sequence of the Neandertal Genome," *Science* 328, no. 5979 (2010): 710–22; B. Wood and N. Lonergan, "The Hominin Fossil Record: Taxa, Grades and Clades," *Journal of Anatomy* 212, no. 4 (2008): 354–376. Below is the full sequence of the human genus from oldest to most recent. The median species longevity is about 500,000 years, but that is uncertain with a spotty fossil record that could be underestimating the longevity of these species and missing various others with narrow distributions or short durations. *Homo habilis* (2.4 million to 1.6 million), *Homo rudolfensis* (2.4 million to 1.8 million), *Homo erectus* (1.9 million to 110,000), *Homo heidelbergensis* (700,000 to 200,000), *Homo neanderthalensis* (400,000 to 40,000), *Homo denisova* (200,000 to 50,000), *Homo floresiensis* (190,000 to 50,000), *Homo sapiens* (315,000 to present).

7   Gregory Stock, *Metaman: The Merging of Humans and Machines into a Global Superorganism* (Simon & Schuster, 1993).

8   The speed 300,000 kilometers/second, fast enough to circle Earth seven times in a blink. A light-year is 5.9 trillion miles, about 63,000 times the distance from the earth to the sun.

9   Planck Collaboration, "Planck 2018 Results. VI. Cosmological Parameters," *Astronomy & Astrophysics* 641 (2020): A6. https://doi.org/10.10 51/0004-6361/201833910. This observation doesn't violate relativity, as space itself is expanding, not objects moving through it.

10   Over infinite time, a spaceship at 1c reaches the cosmic event horizon (17 billion light-years), while at 0.1c, it's limited to ~1.7 billion light-years, due to expansion outpacing slower speeds. Cubing these against the observable universe's radius (63 billion light-years) gives volumes: $(17/63)^3 \approx 0.0197$ (2%) for 1c, and $(1.7/63)^3 \approx 0.0000197$ (0.002%) for 0.1c. These fractions reflect the tiny bubbles each could visit in an ever-stretching cosmos. S. M. Carroll, *Spacetime and Geometry: An Introduction to General Relativity* (Cambridge University Press, 2019).

11   NASA Jet Propulsion Laboratory, "Mars in Our Night Sky: How Far Away Is Mars?" JPL Education (website). Also see https://mars.nasa. gov/all-about-mars/night-sky/close-approach/.

12   Tom Risen, "Selling Mars as Planet B," Aerospace America, May 31, 2017, aerospaceamerica.aiaa.org/selling-mars-as-planet-b/ refers to Stephen Hawking's 2017 comment, "The human species will have to populate a new planet within 100 years if it is to survive."

13   Susan Blackmore projected Darwinian evolution into the digital sphere, where replication, resource competition, and adaptation fuel technological complexity, and she coined the term "temes" (technological memes), which in her 2007 TED talk cast technology as a "third replicator" after genes and memes, shaped by variation, selection, and retention. S. Blackmore, "Memes and 'Temes,'" June 2008, https://www.ted.com/ talks/susan_blackmore_memes_and_temes). For our emergent superorganism, this provides an engine for continuing cognitive evolution when computation pushes into quantum and extra-planetary domains

that are beyond biology's reach. This idea is echoed broadly: Daniel Hillis's vision of a "technological nervous system" imagines tech as a self-sustaining web. W. Daniel Hillis, "The Future of Technology," talk at the GET Conference, Joseph B. Martin Conference Center, Harvard Medical School, Boston, MA, April 26, 2016, video, 10:04, https://www.youtube.com/watch?v=KU2WSQ4XUF4). Kevin Kelly's "Technium" sees tech edging toward autonomy. K. Kelly, *What Technology Wants* (Viking Press, 2010). Stuart Kauffman's "adjacent possible" frames iterative expansion beyond human intent. S. A. Kauffman, *Investigations* (Oxford University Press, 2000). These suggest self-driving processes to propel Metahumanity's cognitive growth indefinitely.

14   David Chalmers's *Reality+* sees uploaded consciousness as authentic, asserting that well-crafted virtual processes carry the same ontological heft as physical ones. D. J. Chalmers, *Reality+: Virtual Worlds and the Problems of Philosophy* (W. W. Norton, 2022). This work portrays uploaded minds as undiminished heirs to their biological origins and pushes to dismantle doubts about non-biological consciousness. He examines how identity might persist through uploading, proposing that such minds could chase "intensive expansion"—deeper, richer experiences—over sprawling "extensive expansion" (378).

15   ChatGPT-3 gave way to ChatGPT-4, then to ChatGPT-4o, DeepSeek R1, Grok 4, Claude 3.7 Sonnet, Gemini 2.0 Pro, and others. And we have only begun. The pace will accelerate, along with eddies of extended persistence for vestigial technologies (and people), in protected nooks.

16   F. J. Dyson, "Search for Artificial Stellar Sources of Infrared Radiation," *Science* 131, no. 3414 (1960): 1667–68, https://doi.org/10.1126/science.131.3414.1667.

17   C. Vidal, "Stellivore Extraterrestrials? Binary Stars as Living Systems," *Acta Astronautica 128* (2016): 251–56, https://doi.org/10.1016/j.actaastro.2016.06.038.

18   The Kardashev scale was developed by astrophysicist Nikolai Kardashev in 1964 to classify civilizations based on the amount of energy they harness and utilize. He defined three types: Type I Civilizations can use all of the energy available on its home planet, roughly estimated at 10^{16} to 10^{17} watts; Type II Civilizations can capture the energy of its host star, for example by constructing megastructures like a Dyson sphere, harnessing on the order of 10^{26} watts; and Type III Civilizations can control energy on the scale of its entire galaxy, which might amount to around 10^{36} watts. N. S. Kardashev, "Transmission of Information by Extraterrestrial Civilizations," *Soviet Astronomy* 8 (1964): 217.

19   The sun's luminosity = 3.82×10^{26} W; energy in 1 ms = 3.83×10^{23} J. Human energy use (2023) $\approx 6.00 \times 10^{20}$ J/year; for 1,000 years = 6.00×10^{23} J. Time to match: $6.00 \times 10^{23} \div 3.83 \times 10^{26} \approx 0.0016$ seconds (1.6 ms). Source: NASA, "Sun Fact Sheet," https://nssdc.gsfc.nasa.gov/planetary/factsheet/sunfact.html; IEA, "World Energy Outlook 2023."

20   It would take light 6.4 minutes to cross a sphere at the orbital distance of Mercury, 12 minutes at that of Venus, and 16.6 minutes at that of Earth. National Aeronautics and Space Administration, "Planetary Fact Sheet," 2023, https://nssdc.gsfc.nasa.gov/planetary/factsheet/

21   Consider a cluster of 1-gram nanodevices, each thinking at 10^{15} operations per second (femtosecond scale), using 10^{-6} watts. A billion such devices (1 kg total) hit 10^{24} ops/sec with just 1 watt. Compare this to a stellivore draining 10^{30} watts from a binary star pair to power a Dyson sphere computer at 10^{20} ops/watt, yielding 10^{50} ops/sec raw—but over 300,000 km, communication lag limits effective throughput to 10^{10} ops/sec. The nanocluster's 10^{24} ops/sec vastly outstrips the stellivore's 10^{10} ops/sec—10^{14} times more compute power—thanks to near-instant 10^{-15} sec communications across micrometers versus stellar-scale delays, showing small can outthink massive with minimal energy.

22   Microsoft's Azure Quantum platform is developing a scalable quantum computer based on topological qubit technology—current prototypes incorporate roughly 10 physical qubits, with a long-term

target of a 1,000-qubit system. Microsoft Corporation, "Azure Quantum," Microsoft Quantum Overview – Quantum Machines | Microsoft Azure, 2023, https://azure.microsoft.com/en-us/solutions/quantum-computing/.

23   X's Grok 3 was trained in late 2024 on a giant computational cluster containing 100,000 Nvidia H100 GPU chips with 80 billion tiny transistors packed into each. PCMag, "Elon Musk's xAI Powers Up 100K Nvidia GPUs to Train Grok," *PCMag*, July 22, 2024, https://www.pcmag.com/news/elon-musk-xai-powers-up-100 k-nvidia-gpus-to-train-grok. NVIDIA's Hopper design under-pinned those chips. NVIDIA Technical Blog, "NVIDIA Hopper Architecture In-Depth," *NVIDIA*, March 22, 2022, https://developer.nvidia.com/blog/nvidia-hopper-architecture-in-depth/. NVIDIA's most powerful chip at the end of 2024 was the Blackwell GPU with 208 billion transistors. AnandTech, "NVIDIA Blackwell Architecture and B200/B100 Accelerators Announced," *AnandTech*, March 18, 2024, https://www.anandtech.com/show/21310/nvidia-blackwell-architecture-and-b200b100-accelerators-announced-going-bigger-with-smaller-data. Grok 4, released in July 2025, had 200,000 GPUs. Jowi Morales, "Musk's Colossus Is Fully Operational with 200,000 GPUs Backed by Tesla Batteries—Phase 2 to Consume 300 MW, Enough to Power 300,000 Homes," *Tom's Hardware*, May 8, 2025, https://www.tomshardware.com/tech-industry/artificial-intelligence/musks-colossus-is-fully-operational-wit h-200-000-gpus-backed-by-tesla-batterie s-phase-2-to-consume-300-mw-enough-to-power-300-000-homes.

24   P. J. Peebles, *Principles of Physical Cosmology* (Princeton University Press, 1993); T. M. Davis and C. H. Lineweaver, "Expanding Confusion: Common Misconceptions of Cosmological Horizons and the. Super-luminal Expansion of the Universe," arXiv, last revised November 13, 2003, https://arxiv.org/abs/astro-ph/0310808.

25   The earliest stars have none of the heavy elements needed to support life. Second-generation stars, here since 3 or 4 billion years after the Big

Bang and formed from gas enriched by supernovae of these first stars (nearly all hydrogen and helium), contain H, He, and a small but significant fraction of heavier elements like C, O, and Fe. Third-generation stars, like the Sun, form from material further enriched by prior stellar deaths, with higher heavy-element content. Planets of first-generation stars have virtually no heavier elements and would be metal-free gas giants. Planets of second-generation stars concentrate limited metals into rocky bodies. Planets, like earth, of third-generation stars have much higher metal levels. Thus, the molecules in your body have been inside of two prior stars. J. L. Johnson and H. Li, "The First Planets: The Critical Metallicity for Planet Formation," *The Astrophysical Journal* 751, no. 2 (2012): 81, https://doi.org/10.1088/0004-637X/751/2/81; V. Bromm and N. Yoshida, "The First Galaxies," *Annual Review of Astronomy and Astrophysics* 49 (2011): 373–407; D. A. Fischer and J. Valenti, "The Planet–Metallicity Correlation," *The Astrophysical Journal* 622, no. 2 (2005): 1102–17.

26   J. Tarter, "The Search for Extraterrestrial Intelligence (SETI)," *Annual Review of Astronomy and Astrophysics* 39 (2001): 511–48.

27   M. H. Hart, "Explanation for the Absence of Extraterrestrials on Earth," *Quarterly Journal of the Royal Astronomical Society* 16 (1975): 128–35.

28   This is less than .01% of the stars in our galaxy, but enough to be interesting. J. Bland-Hawthorn and O. Gerhard, "The Galaxy in Context: Structural, Kinematic, and Integrated Properties," *Annual Review of Astronomy and Astrophysics* 54 (2016): 529–96, https://doi.org/10.1146/annurev-astro-081915-023441.

29   A rudimentary prototype of such a swarm could look at our solar system within a decade. Gregory Stock, "Swarm Evolving Interstellar Reconnaissance (SEIR): A Scalable Vision for Galactic Exploration from a Near-Term Proof of Concept," submitted to *Journal of the British Interplanetary Society* (JBIS), September 2025.

30   C. Vidal, *The Beginning and the End: The Meaning of Life in a Cosmological Perspective* (Springer, 2014), https://doi.org/10.1007/978-3-319-05062-1.

31   Dyson spheres, for example, would show characteristic increased infrared radiation due to the reduced temperatures of the star-encompassing scaffolding. Yet surveys of millions of stars have so far come up empty, turning up only a handful of dubious possibilities for further exploration. M. Suazo, E. Zackrisson, P. K. Mahto, F. Lundell, C. Nettelblad, A. J. Korn et al., "Project Hephaistos – II. Dyson Sphere Candidates from Gaia DR3, 2MASS, and WISE," *Monthly Notices of the Royal Astronomical Society* 531, no. 1 (2024): 695–707, https://doi.org/10.1093/mnras/stae1186.

32   At 20% of the speed of light, such a probe would cross earth's orbit in less than 90 minutes. Low-moving interstellar asteroids do occasionally show up. One, *Oumuamua*, about 200 meters long and traveling at .00002 c, was spotted in 2017. Karen J. Meech, Robert Weryk, Marco Micheli, Jan T. Kleyna, Olivier R. Hainaut, Robert Jedicke et al., "A Brief Visit from a Red and Extremely Elongated Interstellar Asteroid," *Nature* 552, no. 7685 (2017): 378–81, https://doi.org/10.1038/nature25020.

33   The substantial engineering to accomplish it seems feasible for a suitably advanced civilization, but the motivation is unclear. Would an interconnected stellar-level intelligence want to permanently exile a part of itself, condemning it to ten or twenty orders of magnitude less brain power? The loss might not be so significant for a quantum brain, but then again, that quantum brain might find the macro realm a vast and relatively uninteresting desert realm that it departed long ago when it succeeded in unlocking more interesting and connected quantum scales.

34   There is always the possibility, of course, that civilizations have spread through the universe but not in a form that we can recognize. These beings might be like dust or pure energy or in another dimension, or cloaked in some way, or simply beyond our present conceptions. Unknown unknowns will always be present until the universe is fully understood, which may never happen. Thus, we will set aside these pos-

sibilities. They are endless, and my conjectures about the larger human future, even when constrained by our current understanding of the universe's limits, already are almost more than we can grasp.

35   A SETI conversation with a star 500 light-years away might be like a Monty Python routine. Hello . . . *1,000 years pass* . . . *Hello* . . . *1,000 years pass* . . .. *Here is a picture of me* . . . *1,000 years pass* . . . *Thanks, here is one of me.* . . *1,000 years pass* . . . *Here are some prime numbers* . . . *1,000 years pass* . . .*Wow. Cool. You must be pretty smart. Here are some more.* With neither a common language nor communication protocol, a drawn-out conversation like this would be excruciatingly tedious, especially for beings thinking trillions of times faster than we do. And if faster-than-light communication is feasible, then messaging would be via that path, not the radio waves SETI is monitoring.

36   J. F. Kasting, D. P. Whitmire, and R. T. Reynolds, "Habitable Zones around Main Sequence Stars," *Icarus* 101, no. 1 (1993): 108–28, https://doi.org/10.1006/icar.1993.1010.

37   Y. Chen, X.-G. Hou, and H.-L. Lu, "Early Cambrian Vertebrates from Haikou, Kunming, Yunnan, China," *Science* 284, no. 5419 (1999): 1345–48, https://doi.org/10.1126/science.284.5419.1345.

38   NASA Exoplanet Archive, "Confirmed Exoplanets," California Institute of Technology, 2025, https://exoplanetarchive.ipac.caltech.edu.

39   The solar system offers 70–120 celestial bodies, including 8 planets, Ceres, 9 major moons, and 50–100 asteroids with volumes exceeding fifty cubic miles, all suitable for ASI habitats, all within a light-day of one another. These bodies provide access to solar energy (scaled by distance from the Sun) or nuclear power (via radioactive materials like uranium). Planets and moons offer vast volume, while asteroids provide resource-rich, adaptable options. Calculations used planetary diameters and assumed engineering feasibility for hollowing or construction. Details on sizes and resources are available from NASA's Planetary fact

sheets, NASA Solar System Exploration, https://doi.org/10.1006/icar.1993.1010.

40 L. R. Squire and A. J. Dede, "Conscious and Unconscious Memory Systems," *Cold Spring Harbor Perspectives in Biology* 7, no. 3 (2015): a021667, https://doi.org/10.1101/cshperspect.a021667.

41 Clark and Chalmers established the extended mind thesis, arguing that cognition spans into tools and systems when functionally integrated. A. Clark and D. Chalmers, "The Extended Mind," *Analysis* 58, no. 1 (1998): 7–19, https://doi.org/10.1093/analys/58.1.7. Paul Smart went on to link cognition to computational webs like the internet, where digital networks amplify human capacity. P. R. Smart, "Extended Cognition and the Internet: A Review of Current Issues and Controversies," *Philosophy & Technology* 30, no. 3 (2017): 357–90. Their work suggests that non-biological elements—code, circuits, clouds—become intrinsic to mind, a trajectory that leads toward the global mind of Metahumanity.

42 For example, Hugh Herr's sensory-augmented lower legs, as in H. M. Herr and T. R. Clites, "Agonist-Antagonist Myoneural Interface for Functional Limb Restoration After Amputation," *Journal of Orthopaedic Research* 36, no. 11 (2018): 2921–29, https://doi.org/10.1002/jor.24100.

43 M. Pollan, *How to Change Your Mind: What the New Science of Psychedelics Teaches Us About Consciousness, Dying, Addiction, Depression, and Transcendence* (Penguin Press, 2018); Easwaran, *The Upanishads*, 2nd ed. (Nilgiri Press, 2007).

44 L. Margulis and D. Sagan, *What Is Life?* (University of California Press, 1995).

45 Melody R. Conklin, "Cat Whiskers 101," *Zoetis Petcare* (blog), accessed June 29, 2025, https://www.zoetispetcare.com/blog/article/cat-whiskers-101.

46   E. Ruppert, R. S. Fox, and R. D. Barnes, *Invertebrate Zoology: A Functional Evolutionary Approach*, 7th ed. (Brooks/Cole, 2004).

47   These sensory swarms also might provide a long warning of everything in the vicinity. An intruder 100 light-years away would take 400 years to reach us at 0.2c.

48   M. Tegmark, *Life 3.0: Being Human in the Age of Artificial Intelligence* (Knopf, 2017).

49   B. Greene, *The Hidden Reality: Parallel Universes and the Deep Laws of the Cosmos* (Knopf, 2011).

50   N. J. Poplawski, "Cosmology with Torsion: An Alternative to Cosmic Inflation," *Physics Letters B* 694, no. 3 (2010): 181–85, https://doi.org/10.1016/j.physletb.2010.09.056.

51   G. Tononi and C. Koch, "Consciousness: Here, There and Everywhere?" *Philosophical Transactions of the Royal Society B: Biological Sciences* 370, no. 1668 (2015): 20140167, https://doi.org/10.1098/rstb.2014.0167.

52   Kevin Kelly's insights in *What Technology Wants* depict technological evolution as a force trending toward greater complexity and consciousness. K. Kelly, *What Technology Wants* (Viking Press, 2010). This reality resonates with Pierre Teilhard de Chardin's "noosphere," a collective consciousness nearing an "Omega Point" of ultimate unity, though de Chardin casts it in biological and spiritual terms and sees it as steered by divine will. P. Teilhard de Chardin, *The Phenomenon of Man* (Harper & Brothers, 1959). The universe's vastness and light-speed limits, however, add a wrinkle by suggesting that any cosmic design would spawn not one Omega Point but countless localized ones, which complicates de Chardin's grand vision. This leaves a stark tension: Does macro-evolution—biological and technological—trace a purposeful and spiritual arc, or merely reflect emergent properties of a complex, material universe? For Metahumanity, resolving this by discerning what,

if anything, lies behind the universe's laws and structure may stand as the ultimate challenge or be relatively accessible.

53   V. E. Frankl, *Man's Search for Meaning* (Beacon Press, 1959).

54   I've mentioned that competition might fuel Metahumanity's relentless push for compute. Reproductive variation and differential selection drive biological evolution, and Metahumanity's subsystems, nodes, networks, and processes vary endlessly. The more adept ones don't just endure; they spread, replicating and displacing what lags. This dance of aggressive variation and selection digitally echoes our own evolution but is accelerated beyond biology's glacial pace. From this vantage point, compute scales because its contributors must in order to survive.

55   Dyson spheres are shells or swarms harnessing a star's 10^{26} watts, 15 orders of magnitude more than all humanity consumes today, and enough to run a computational behemoth. Its mass ($\sim 10^{18}$ kg) and light-speed lags (17 minutes to cross its diameter), however, would bog down communication across such a sphere, even if its mass could be purloined from asteroids and planets. Such lags would make it less efficient than small-scale tech. Freeman J. Dyson, "Search for Artificial Stellar Sources of Infrared Radiation," *Science* 131, no. 3414 (1960): 1667–68, DOI: 10.1126/science.131.3414.1667.

56   R. Kurzweil, *The Singularity Is Nearer: When We Merge with AI* (Viking, 2021). A trillion specks (10^{12}) at 10^{24} FLOPS each—scaling from 10^{10} FLOPS in 2025 via quantum nanotech—yields 10^{36} FLOPS, a speculative peak by 2045.

57   Nanodust: specks (10^{-12} kg) collectively reaching 10^{30} FLOPS on $\sim 10^{20}$ watts, with lags of 10^{-15} seconds. Manufacturing requires vast infrastructure—cryogenic systems, precision factories—despite the swarm's featherweight total. S. Lloyd, "Ultimate Physical Limits," *Nature* 406 (2000): 1047–54.

58   John Smart suggests that sufficiently advanced civilizations do not expand outward into space indefinitely, but instead evolve inward toward denser, more efficient, and ultimately nanoscale or black hole–like computational environments, effectively "leaving" our observable universe. J. M. Smart, "The Transcension Hypothesis: Sufficiently Advanced Civilizations Invariably Leave Our Universe," *Acta Astronautica* 78 (2012): 55–68, https://doi.org/10.1016/j.actaastro.2011.11.006.

59   If nanodust stumbles, Metahumanity will have to adapt by blending small-scale swarms with mid-tier solutions like fusion reactors or solar grids—a pragmatic dance that would push compute larger and smaller, embracing whatever works. W. H. Zurek, "Decoherence, Einselection, and the Quantum Origins of the Classical," *Reviews of Modern Physics* 75, no. 3 (2003): 715–75, https://doi.org/10.1103/RevModPhys.75.715.

60   Metahumanity's potential computational power, a billion billion billion (10^{27}) times the 8×10^{24} operations of all human minds (10^{15} ops/brain for 8 billion people), reaches 8×10^{51} ops. Nanodust swarms (10^{20} specks at 10^{30} ops) scaled to 10^{21} units could achieve this with 10^{29} watts and nanosecond lags, suggesting a decentralized, efficient superorganism. Stock, *Metaman*; K. E. Drexler, *Nanosystems: Molecular Machinery, Manufacturing, and Computation* (John Wiley & Sons, 1992).

61   Physics might cap nanodust at 10^{15} FLOPS per watt before noise drowns the signal, or energy scarcity might halt growth. M. A. Nielsen and I. L. Chuang, *Quantum Computation and Quantum Information* (Cambridge University Press, 2010).

62   The "Flourishing Study" is an interesting look at the search for meaning and purpose across cultures. Eric Kim et al., "Identifying Childhood Correlates of Adult Purpose and Meaning Across 22 Countries (Global Flourishing Study)," *npj Mental Health Research* 4, no. 1 (2025): 14, https://doi.org/10.1038/s44184-025-00127-9; Gallup, "The Global Flourishing Study: Exploring the Science and Meaning of Human Flourishing," YouTube video, 36:42, streamed April 30, 2025, https://www.youtube.com/watch?v=HFoIbJceY7E.

63 *Holy Bible*, New Revised Standard Version, ed. Thomas Nelson (Melton, 1989). Genesis 1:31: "God saw everything that he had made, and indeed, it was very good."

64 *Holy Bible*, New Revised Standard Version, ed. Thomas Nelson (Melton, 1989). John 3:16: "For God so loved the world that he gave his only Son, so that everyone who believes in him may not perish but may have eternal life."

65 *The Qur'an*, trans. M. A. S. Abdel Haleem (Oxford University Press, 2004). Surah 51 (Adh-Dhariyat), Ayah 56: "I created jinn and mankind only to worship Me."

66 *Bhagavad Gita*, trans. Eknath Easwaran (Nilgiri Press, 2007). Chapter 4, verse 8: "To protect the righteous, to destroy the wicked, and to establish the kingdom of God, I come into being age after age."

67 Why order rather than chaos? Why love rather than solitude? Why worship, when God needs no echo? Why play, when stillness would suffice?

68 Friedrich Nietzsche, *The Gay Science: With a Prelude in Rhymes and an Appendix of Songs*, trans. Walter Kaufmann (Vintage Books, 1974).

69 Jean-Paul Sartre agreed. No essence sets us; we choose love, despair, purpose, bearing the weight alone. Jean-Paul Sartre, *Being and Nothingness: An Essay on Phenomenological Ontology*, trans. Hazel Barnes (Philosophical Library, 1943).

70 Ilia Delio, a Franciscan sister trained in neuroscience and deeply influenced by Teilhard de Chardin, has extended his vision of spiritual evolution into the age of AI. She argues that rising complexity and consciousness—now manifesting in technological systems—may reflect an emergent divine reality, drawing humanity toward deeper spiritual integration and cosmic wholeness. Ilia Delio, *The Unbearable Wholeness of Being: God, Evolution, and the Power of Love* (Orbis Books, 2013). See also

P. Teilhard de Chardin, *The Phenomenon of Man,* trans. Bernard Wall (Harper & Brothers, 1955).

71   Albert Camus, *The Myth of Sisyphus*, trans. Justin O'Brien (Hamish Hamilton, 1942).

Afterword: A Strategy of Adaptive Adoption

1   Mustafa Suleyman, *The Coming Wave: Technology, Power, and the Twenty-First Century's Greatest Dilemma* (Crown, 2023).

2   Future of Life Institute, "Pause Giant AI Experiments: An Open Letter," March 29, 2023, futureoflife.org/open-letter/pause-giant-ai-experiments/.

3   Ray Wu, "AI Is Rebuilding American Manufacturing," *GenAI Works Newsletter*, April 29, 2025, https://newsletter.genai.works/p/ai-is-rebuilding-american-manufacturing.

4   Kylie Robison, "Mark Zuckerberg's AI Recruiting Spree Finds a New Target," *WIRED*, July 29, 2025, https://www.wired.com/story/mark-zuckerberg-ai-recruiting-spree-thinking-machines/.

5   Grok 4, released in July 2025 had 200,000 GPUs. Jowi Morales, "Musk's Colossus Is Fully Operational with 200,000 GPUs Backed by Tesla Batteries—Phase 2 to Consume 300 MW, Enough to Power 300,000 Homes," *Tom's Hardware*, May 8, 2025.

6   Dylan Butts, "U.S. Chip Controls Will Benefit China's Nvidia Rivals Like Huawei: Analysts," *CNBC*, April 21, 2025, https://www.cnbc.com/2025/04/21/us-chip-controls-boon-for-china-nvidia-rivals-like-huawei-analysts-.html.

7   Suleyman, *Coming Wave.*

8   Max Tegmark, *Life 3.0: Being Human in the Age of Artificial Intelligence* (Knopf, 2017).

9   Stuart Russell, *Human Compatible: Artificial Intelligence and the Problem of Control* (Viking, 2019).

10   Gary F. Marcus, *Taming Silicon Valley: How We Can Ensure That AI Works for Us* (MIT Press, 2024).

11   Roman V. Yampolskiy, *AI: Unexplainable, Unpredictable, Uncontrollable* (Routledge, 2024).

12   Eliezer Yudkowsky and Nate Soares, *If Anyone Builds It, Everyone Dies: Why Superhuman AI Would Kill Us All* (Little, Brown and Company, 2025).

13   Belief that ASI is not only an existential threat to humanity, but one we could parry with the right policies and vigilance, leads to the conclusion that no cost is too high to save humankind. This mindset is built on both fear of the future and the conceit that we can predict and preempt that future. Those seeking to prevent the emergence of AGI imagine themselves guardians of humanity, but so did history's worst tyrannies. In Stalin's Russia, Mao's China, and Hitler's Germany, the architects of repression and slaughter believed they were protecting humankind from mortal threats. To save humanity by dismantling the freedoms we cherish would annihilate what we have built—and still end in failure. AI's trajectory cannot be frozen, nor should it be. The wiser course is humility in our predictions, alertness to emerging challenges, flexibility in our responses, hope about coming possibilities, and courage in accepting a future of hybrid cognition within Metahumanity.

14   If an AI is operating a million times faster than we are, then 3 months of human time equals 1 sec of AI time. We are like trees to them, so slow and stationary, that it would take time-lapse to see us move. Good luck at controlling them, given that decades of AI time would pass before we could even articulately discuss the issue.

15   For some intriguing perspectives on this, see John M. Smart, Overview, *Natural Alignment* (newsletter), Substack, 2002, https://naturalalignment.substack.com/.

16   It is a mistake to think that limiting AI's understanding of humanity and the world will make us safer. If human/AI alignment already exists, then helping AI fully understand us adds to our safety.

17   Edward O. Wilson, quoted in "An Intellectual Entente," *Harvard Magazine,* September 10, 2009, last updated November 24, 2020.

18   See Anshad Ameenza, "Beyond LLMs: The Uncharted Path to Artificial General Intelligence," March 30, 2025, https://anshadameenza.com/blog/technology/beyond-llms-path-to-agi/.

19   Here is a detailed overview of current US policy objectives and scope: Executive Office of the President, *America's AI Action Plan*, Washington, DC: The White House, July 2025, accessed July 28, 2025, https://www.whitehouse.gov/wp-content/uploads/2025/07/Americas-AI-Action-Plan.pdf. The authors embrace the criticality of aggressive AI development, push aggressively to moderate regulation, argue for expanded energy production, and want to cement US leadership in AI. They believe that rapid AI adoption and advancement provide the surest path to US security and prosperity, and they want to empower the private sector to move ahead rapidly in this effort. This document speaks of a *race-to-win* and is very bullish about the potential for aggressive policy initiatives to move the United States forward rapidly. This competitive framing seems to match similar sentiments in China. While the EU and various policy critics may voice a preference for more caution, the current landscape is far from a foundation for measured, cautious development of AI. This is a race to who knows where.

20   Kathryn A. Glatter and Paul Finkelman, "History of the Plague: An Ancient Pandemic for the Age of COVID-19," *The American Journal of Medicine* 134, no. 2 (2020): 176–81, https://pmc.ncbi.nlm.nih.gov/articles/PMC7513766/.

21 These systems represent the essential biological machinery: (1) DNA replication ensures accurate genetic inheritance; (2) transcription and translation produce proteins from genes; (3) metabolism enables energy transformation; (4) cell division governs growth and development; (5) homeostasis maintains internal stability; (6) cell signaling coordinates function; (7) cellular repair prevents damage accumulation; and (8) reproduction enables continuity of the species.

22 Gillian Bejerano et al., "Ultraconserved Elements in the Human Genome," *Science* 304, no. 5675 (2004): 1321–25, https://doi.org/10.1126/science.1098119; Solomon Katzman et al., "Human Genome Ultraconserved Elements Are Ultraselected," *Science* 317, no. 5840 (2007): 915, https://doi.org/10.1126/science.1142430 These sources show approximately 20-fold lower mutation rates in these regions compared to genome averages.

23 R. Linnakaari, N. Helle, M. Mentula, A. Bloigu, M. Gissler, O. Heikinheimo, and M. Niinimäki, "Trends in the Incidence, Rate and Treatment of Miscarriage — Nationwide Register-Study in Finland, 1998–2016," *Human Reproduction* 34, no. 11 (2019): 2120–28, https://doi.org/10.1093/humrep/dez211 This source estimates total embryo loss of >30%, largely due to chromosomal anomalies or disruptions in core developmental functions.

24 The control component is significant. If we have access to avatars that we grow emotionally attached to as they become our trusted guides and companions, it would be devastating to lose access. There would be a host of potential problems unless the avatar is local, private, and protected. Could the price of access be raised? Could our most intimate moments be discovered by others? Could a model be discontinued so that we lose our friend? These issues are not trivial and not easily solved. Peter Voss captures them nicely in his concept of the *Personal Personal Assistant*—adapted to us individually, owned by us individually, under our personal control, and private. See Peter Voss, "Imagine," *Peter's Substack*, April 9, 2024, https://petervoss.substack.com/p/imagine. Also, "All I Wanted Was an Option to Keep 4o," Reddit forum

r/ChatGPT, August 8, 2025, https://www.reddit.com/r/ChatGPT/comments/1mkhfep/all_i_wanted_was_an_option_to_keep_4o/. This reddit forum discusses the devastation that users were feeling from OpenAI's sudden discontinuation of ChatGPT-4o. The nuances of exchanges that users had been having for more than a year suddenly shifted with ChatGPT-5, and users lost a confidant and understanding voice. The backlash was so strong that OpenAI turned the model back on. This will be a major issue. Imagine how much more profound the loss would be if the relationships extended over a decade and the AI personas were true companions. Even minor changes might make you feel that your "friend" was irrevocably lost, as the human brain has evolved to be extremely sensitive to shifts in mood and presence. Or for that matter, what would happen if you and your sophisticated avatar grew apart—just like people do—and no longer liked each other?

25   Peter G. Peterson Foundation, "Almost 25% of Healthcare Spending is Considered Wasteful. Here's Why," April 3, 2023, https://www.pgpf.org/article/almost-25-percent-of-healthcare-spending-is-considered-wasteful-heres-why.

26   The *precautionary principle* holds that when an action carries a plausible risk of serious harm, society should err on the side of caution—even if scientific evidence is not yet conclusive—by shifting the burden of proof onto those proposing potentially harmful activity, which can be nearly impossible to obtain. This approach elevates preventive regulation over reactive fixes, and the principle has been widely invoked by environmental advocates to justify moratoria on everything from genetically modified crops to large-scale infrastructure projects, a stance critics argue can unduly slow or halt technological and economic development. See Gregory Conko, "Throwing Precaution to the Wind: The Perils of the Precautionary Principle," *Competitive Enterprise Institute*, September 1, 2000, https://cei.org/publication/throwing-precaution-to-the-wind-the-perils-of-the-precautionary-principle.

27   Thucydides, *History of the Peloponnesian War*, Book 2, Section 40, trans. Richard Crawley, 1874. Available in various editions, such as the Penguin Classics version (1972, trans. Rex Warner) or online at Perseus Digital Library.